Mathematical Tools for
Shape Analysis and Description

Synthesis Lectures on Computer Graphics and Animation

Editor
Brian A. Barsky, *University of California, Berkeley*

This series will present lectures on research and development in computer graphics and geometric modeling for an audience of professional developers, researchers and advanced students. Topics of interest include Animation, Visualization, Special Effects, Game design, Image techniques, Computational Geometry, Modeling, Rendering and others of interest to the graphics system developer or researcher.

Mathematical Tools for Shape Analysis and Description

Silvia Biasotti, Bianca Falcidieno, Daniela Giorgi, and Michela Spagnuolo

ISBN:978-3-031-79557-2   paperback
ISBN:978-3-031-79558-9   ebook

DOI 10.1007/978-3-031-79558-9

A Publication in the Springer series
SYNTHESIS LECTURES ON COMPUTER GRAPHICS AND ANIMATION

Lecture #16
Series Editor: Brian A. Barsky, *University of California, Berkeley*
Series ISSN
Print 1933-8996   Electronic 1933-9003

Mathematical Tools for Shape Analysis and Description

Silvia Biasotti, Bianca Falcidieno, Daniela Giorgi, and Michela Spagnuolo
Consiglio Nazionale delle Ricerche, Italy

SYNTHESIS LECTURES ON COMPUTER GRAPHICS AND ANIMATION #16

ABSTRACT

This book is a guide for researchers and practitioners to the new frontiers of 3D shape analysis and the complex mathematical tools most methods rely on. The target reader includes students, researchers and professionals with an undergraduate mathematics background, who wish to understand the mathematics behind shape analysis.

The authors begin with a quick review of basic concepts in geometry, topology, differential geometry, and proceed to advanced notions of algebraic topology, always keeping an eye on the application of the theory, through examples of shape analysis methods such as 3D segmentation, correspondence, and retrieval.

A number of research solutions in the field come from advances in pure and applied mathematics, as well as from the re-reading of classical theories and their adaptation to the discrete setting. In a world where disciplines (fortunately) have blurred boundaries, the authors believe that this guide will help to bridge the distance between theory and practice.

KEYWORDS

computational topology, differential geometry, algebraic topology, spectral methods, shape invariants, distance measures, shape transformations, 3D shape analysis, 3D shape description, 3D shape retrieval, Morse theory, topological persistence

Contents

Acknowledgments

The National Research Council (Consiglio Nazionale delle Ricerche, CNR) of Italy supported the research activity that originated much of the material presented in this book. We also acknowledge the support of projects that contributed to the development of infrastructures for 3D data sharing, in particular AIM@SHAPE, FOCUS K3D and the ongoing FP7 INFRASTRUCTURE VISIONAIR.

We wish to thank all our co-authors who have worked with us on shape analysis and description, whether or not related to the areas that we specifically address. In particular, we mention Andrea Cerri, Massimo Ferri, Patrizio Frosini, Claudia Landi, Giuseppe Patanè, Michela Mortara, Marco Attene, Chiara Catalano and all members of the Shape Modeling Group at CNR IMATI, for the helpful discussions and contributions over the years. Daniela Giorgi also wishes to thank Ovidio Salvetti and colleagues from the Signals and Images Lab at CNR ISTI for their precious support.

The wise suggestions and comments of the book reviewers were greatly appreciated, and we hope they are reflected in better content, coverage and structure of this work. Thanks are also due to our publisher, Michael Morgan, for his constant encouragement and patience over our ever-delayed deadlines.

Silvia Biasotti, Bianca Falcidieno, Daniela Giorgi, and Michela Spagnuolo
August 2014

Figure Credits

Figure 1.1   from [195], D. W. Thompson, editor. On Growth and Form. Copyright © 1942, Cambridge University Press, Cambridge, Ma. 1942. Used with permission.

Figure 4.7   from [145], M. Mortara, G. Patané, M. Spagnuolo, B. Falcidieno, and J. Rossignac. Blowing bubbles for multi-scale analysis and decomposition of triangle meshes. *Algorithmica*, 38(2):227–248, 2003. Copyright © 2003, Springer-Verlag. Used with the kind permission of Springer Science+Business Media.

Figure 5.1/5.2   from [169], M. Reuter, S. Biasotti, D. Giorgi, G. Patané, and M. Spagnuolo. Discrete Laplace-Beltrami operators for shape analysis and segmentation. Comput. Graph., 33(3):381–390, June 2009. Copyright © 2009 Elsevier. Used with permission.

Figure 5.3   from [125], B. Levy. Laplace-Beltrami eigenfunctions towards an algorithm that "understands" geometry. In Proceedings of the IEEE International Conference on Shape Modeling and Applications 2006, SMI '06, pages 13–, Washington, DC, USA, 2006. IEEE Computer Society. Copyright ©2006 IEEE. Used with permission.

Figure 6.5/6.6   from [128], Y. Lipman and T. Funkhouser. Möbius voting for surface correspondence. ACM Trans. Graph., 28(3):72:1–72:12, July 2009. Copyright © 2009, Association for Computing Machinery. Reprinted by permission.

Figure 7.6   from [59], T. K. Dey, A. N. Hirani, and B. Krishnamoorthy. Optimal homologous cycles, total unimodularity, and linear programming. In Proceedings of the 42nd ACM symposium on Theory of computing, STOC '10, pages 221–230, New York, NY, USA, 2010. Copyright © 2010, Association for Computing Machinery. Reprinted by permission.

Figure 9.4   from [197], J. Tierny, J.-P. Vandeborre, and M. Daoudi. Partial 3D shape retrieval by Reeb pattern unfolding. Comput. Graph. Forum, 28(1):41–55, 2009. Copyright © 2009, Eurographics. Used with permission.

Figure 10.3   from [33], P. T. Bremer and V. Pascucci. A practical approach to two-dimensional scalar topology. In H. Hauser, H. Hagen, and H. Theisel, editors, Topology-based Methods in Visualization, pages 151–169. Springer Berlin Heidelberg, 2007. Copyright © 2007, Springer. Used with the kind permission of Springer Science+business Media.

Figure 12.3/ 12.4   from [106], S.-S. Huang, A. Shamir, C.-H. Shen, H. Zhang, A. Sheffer, S.-M. Hu, and D. Cohen- Or. Qualitative organization of collections of shapes via quartet analysis. ACM Transactions on Graphics, 32(4):71:1–71:10, July 2013. Copyright © 2013, Association for Computing Machinery. Reprinted by permission.

Figure 13.1   The Coseg Main Page. Courtesy of Yunhai Wang. Used with permission.

CHAPTER 1

About this Book

1.1   SHAPE AND SHAPE ANALYSIS

Modeling shapes is part of both cognitive and creative processes, and from the outset, models of physical shapes have satisfied the desire to see in advance the result of a project. For over a century philosophers and psychologists have tried to understand how the human visual perception system recognizes objects from images as picked up from the human eye. Sensory input and perceived shape geometry are two of the key elements, but not the only, which influence humans' understanding of objects. Prior knowledge about an object's place and functional context, perceived similarity to other presented objects, or current task demands, can equally contribute to an object description.

Shape analysis may be defined as a set of theories, methods and algorithms that concur to the formalization and computation of properties useful to characterize the geometrical appearance of objects. Initiated by computer vision and pattern recognition, the focus of shape analysis was on the identification of objects in images. A large number of methods was developed, relying on experimentation on how humans' representations of objects are built bottom-up from low-level 2D image features. Object recognition and reconstruction are particularly challenging in vision because 2D images contain only partial information about objects due to occlusions and lighting conditions.

Over the last decades, advances in acquisition and modeling techniques made 3D objects as common as images and videos. With the advancements of geometric modeling and computer graphics, shape analysis of 3D digital representations of objects is now maturing to a key discipline penetrating many applied domains, spanning from entertainment to life sciences, from cultural heritage to industrial design, from manufacturing to urban planning. Over the years, computer graphics started addressing the same basic issues targeted by computer vision, from shape recognition to shape segmentation and understanding. Yet, new techniques had to be developed, as 3D models are different from 2D images: they portrait the complexity of the 3D world, whereas they do not suffer from the sensory gap. Objects represented by 3D models can be analyzed relying on a complete digital model of their shape, and therefore it is possible to evaluate them not only using local low-level features, but also global high-level geometric and topological properties.

In the last years, we observed an explosion of methods for the analysis of 3D shapes and shape collections, targeting a variety of tasks, such as shape design and synthesis, content-based retrieval, and automatic classification. This suggests how 3D shape analysis is likely to become fundamental in the development of future interfaces with the digital worlds. Emerging new technologies are causing a fast process of dematerialization in our lives, with a reduction of the material essence of our reality: material objects are gradually substituted by processes and services which are more and more immaterial [132]. While attaining visual realism is within the grasp of current research and development, further research has to be done to ensure that digital environments will be accommodating to the needs of human communication and interaction. Shape analysis methods are defining the building blocks of these future interfaces, laying the basis for associating meaning to digital assets and for interacting with the digital world in a semantically rich modality.

1.2   WHY *MATH* FOR 3D SHAPE ANALYSIS?

The focus of this book is on *mathematics* for shape analysis: the motivation for this choice is captured very well by the following citation from the famous book *On Growth and Form* by D'Arcy Thompson [195].

> "We must learn from the mathematician to eliminate and discard; to keep in mind the type and leave the single case, with all its accidents, alone; and to find in this sacrifice of what matters little and conservation of what matters much one of the peculiar excellences of the method of mathematics."

In the context of biological taxonomy, D'Arcy Thompson clearly states that mathematical formalisms and tools are needed in shape analysis and classification for their abstraction and synthesis power. He clarifies further: "the study of form may be descriptive merely, or it may become analytical. We begin by describing the shape of an object in the simple words of common speech: we end by defining it in the precise language of mathematics: and the one method tends to follow the other in strict scientific order and historical continuity." Based on this approach, he classifies animals, bones or plants by using geometric transformations (combinations of dilations and contractions). A famous example taken from his book shows two different fish made equivalent through such a transformation (Figure 1.1).

The notion of shape and invariance occurs frequently in many attempts to clarify what a *shape* is and what are the *important* properties to characterize it. In the field of statistical shape analysis, Kendall [111] defined a shape as "all the geometrical information that remains when location, scale, and rotational effects (Euclidean transformations) are filtered out from an object"; yet, if we look at the examples in Figure 1.2, we will probably think there should be something more beside Euclidean transformations. The cups' shapes are indeed similar, even if not all of them may be obtained by Euclidean transformations applied to the same geometry.

The description of shapes relies on the use of similes, as it emerges nicely in the following linguistic definition: "shape: the outer form of something by which it can be seen (or felt) to be different from something else" [3]. This definition establishes a link between the concept of *shape* and the concept of *similarity*, as used for distinguishing among objects: shape emerges thanks to invariance, or equivalence, under similarity transformations. Though convincing, this idea can be puzzling at the same time, as long as we do not define what similar, or different, means. In

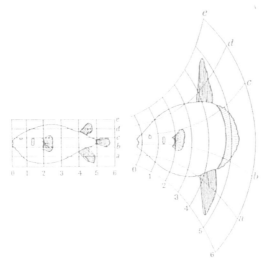

Figure 1.1: Transformation between similar biological shapes [195].

Figure 1.2: Can we transform one cup into another by translation, scaling and rotations? [217]

Figure 1.3, the objects are perceived easily as similar by our brain: they are all pots. The criterion we use is based on a known object's functionality and the perceived similarity follows this knowledge. However, the objects show rather different geometries and textures, though their structure is similar.

Probably, we have to stand with Koenderink [117], who observed that "things possess a shape for the observer, in whose mind the association between the perception and the existing conceptual models takes place." In other words, shape analysis and understanding are cognitive processes, and as such they are dependent upon the context, and consequently the viewpoint in observing shapes.

Figure 1.3: What do *similar* and *different* mean?

These considerations suggest a more operational definition, where the *shape* is a property of *both* a set of objects and a particular method of observation, or analysis. Starting from this assumption, the mathematics discussed in this book is related to approaches which can be used to define the universe of objects and the methods of observation. Spaces, metrics, invariants and transformations are presented at the theoretical and application level, to explain how to derive shape information from objects stored in the form of geometric models. The book cannot cover the wealth of mathematical knowledge pertaining to 3D shape analysis, but it attempts to group recent, and rather complex mathematics, which is necessary to understand and apply techniques at the state-of-the-art. With this book readers will familiarize themselves with basic concepts of geometry and topology, then proceed to more advanced concepts in differential geometry and topology, up to algebraic topology.

1.3   WHAT THIS BOOK IS AND WHAT IT IS NOT

The book is meant to be:

- A guide to the mathematics needed for 3D shape analysis, ranging from concepts in differential geometry to notions of algebraic topology;

- An introductory reading to problems, solutions, and applications of mathematics for 3D shape analysis;

- A window into the literature on 3D shape analysis dealing with spaces, metrics, invariants and transformations;

- A bridge to mathematical textbooks where additional information about theoretical details can be found;

- A pointer to resources for 3D shape analysis, including websites, software, data repositories.

This book was conceived as a *synthetic* guide to mathematics for shape analysis: it is neither meant to be a comprehensive book on every kind of mathematics useful for shape analysis, nor an exhaustive review of the literature in the field. The methods selected from the literature serve as examples, to show the usage context of the mathematics described and its usefulness in shape analysis.

We do not cover important aspects of mathematics: for instance, approximation theory, algebraic geometry, mathematical morphology are not covered by the book. The selection of mathematical topics has been imposed by space limits, with no implication that what has not been covered is unimportant: the choice has been guided by our feeling as practitioners, and we believe the material collected in this book is enough to enable readers to search and find elsewhere specific topics which have been possibly left out. Also, this text does not cover, if not for some examples, the very important problem of the discretization of the mathematical concepts to the domain of 3D *digital* shapes. There are very good and exhaustive books, tutorials and teaching material that fully cover this part of the discussion about mathematics and shape analysis. The Discrete Exterior Calculus, DEC for short, is the right keyword to find reference material in this area (see also [1, 103]). Finally, we mention algorithms that implement the techniques described without digging into the technical details. We believe that knowledge of the underlying mathematics is extremely important to devise a correct implementation of shape analysis techniques. We hope that, after going through the book, the reader will have a better perspective on the mathematical framework that lies behind an algorithm.

1.4   EXPECTED READERS

This book is intended for an audience having an undergraduate mathematics background, for instance:

- University students (graduate or postgraduate) in computer science, engineering, architecture, medicine, bioinformatics, information technology and communication, mathematics, physics;

- Researchers, professionals with a background in mathematics, in particular basic notions of calculus, geometry, algebra and topology;

- Teachers and students of professional training courses;

- Researchers and academics interested in learning the mathematical fundamentals of 3D shape analysis;

- Professional scientists in industry who wish to understand, implement and use 3D shape analysis techniques in their own specific field.

1.5   HOW THIS BOOK IS ORGANIZED

In chapter 2, *shape analysis in a nutshell*, we give an overview of shape analysis discussing problems and application areas. We move then to the technical part, whose structure reflects our attempt to synthesize the presentation into coherent groups of mathematical concepts, where the driving criteria is the perspective on shape analysis. When the shape is analyzed with a geometrical point of view, we are interested in measuring properties, finding characterizations of features, e.g., protrusions, possibly at different scales and resolutions. Geometry and topology provide necessary background to build three kinds of descriptions (see chapter 3), which are elaborated in chapter 4 where concepts of differential geometry are discussed together with examples of applications. Spectral methods applied to 3D meshes are introduced in chapter 5 to show how the multi-resolution study of shapes may be addressed by the extraction and analysis of the *eigen-structures*, as it is done for signal processing. From the analysis of a single shape, we move forward to the analysis of pairs of shapes, by introducing concepts pertaining to the transformations between shapes in chapter 6; the chapter is rather theoretical and introduces a number of interesting distances between spaces.

We skim through algebraic topology in chapter 7 to land on differential topology in chapter 8, where a consistent bunch of mathematics is presented. These concepts are needed to understand a set of methodologies for analyzing shapes which are characterized by the usage of real-valued functions to detect important features of the shape: Reeb graphs in chapter 9, Morse complexes in chapter 10 and topological persistence in chapter 11.

In chapter 12 we introduce a number of concepts and methods where objects are not seen solely as geometric shapes, but also as objects with colors or textures, objects with a functional meaning, or objects that live in collections of many other models. We conclude the book with chapter 13 with an overview of existing resources in the shape analysis field.

In all chapters, besides chapter 3, which is mostly theoretical and 13, we have included one or more sections, named *concepts in action*, which discusses applications of the mathematics defined.

The background knowledge in mathematics that we rely on concerns:

- basic elements of analysis: e.g., functions, continuity, differentiability;

- basic elements of linear algebra: e.g., vectors, matrices, eigenvalues;

- basic elements of set theory and algebra: e.g., sets, groups.

References to specific textbooks where readers may find additional information on the mathematics described are given in the chapters where they are expected to be more useful.

C H A P T E R   2

3D Shape Analysis in a Nutshell

2.1   3D SHAPE ANALYSIS: PROBLEMS AND SOLUTIONS

Computer graphics and geometric modeling emerged as new disciplines focusing on the design and visualization of digital 3D objects. At the beginning, they mostly focused on solving primary problems related to representation issues [134, 167]; the seminal paper by Requicha [167] introduced the basic terminology and definitions which shaped the whole field of geometric modeling. Then, computer graphics saw a gradual shift of interest from methods to *represent* shapes toward methods to *describe* shapes. The distinction between *representation* and *description* can be expressed as follows [150]:

> "an object representation contains enough information to reconstruct (an approximation to) the object, while a description only contains enough information to identify an object as a member of some class."

In other words, the representation of an object is more detailed and accurate than a description, but it does not necessarily contain any high-level information on the object explicitly. The description is more concise, still it conveys an elaborate and composite view of the object.

Over the years, computer graphics started addressing the same basic issues targeted by computer vision, from shape recognition to shape segmentation and understanding. The analysis of 3D models is completely different from the analysis of 2D images. The advantage in 3D is that we rely on a complete representation of the object shape, while the prospective projection in 2D may cause gaps due to occlusions or to lighting conditions. The disadvantage in 3D is that geometric models and representations are much more complex to handle than simple grids of pixels. Nowadays, 3D shape analysis deals with an ever-growing number of challenging issues, from the analysis of single complex shapes (e.g., deformable models), to the reasoning on the content of entire collections of shapes. The range of problems pertaining to 3D shape analysis is reflected by the list of research topics below (see also Figure 2.1); throughout the book we will see how mathematics can support the solution of these problems.

3D shape analysis problems

- **Feature detection: How can we find significant geometric features robustly?** Feature detection refers to the identification of significant shape features, where the term "significant" has a different meaning in different domains. The automatic detection of surface features is a fundamental task in 3D shape analysis, as it facilitates operations such as editing, simplification, morphing, and compression of digital 3D models. Also, feature detection is often the

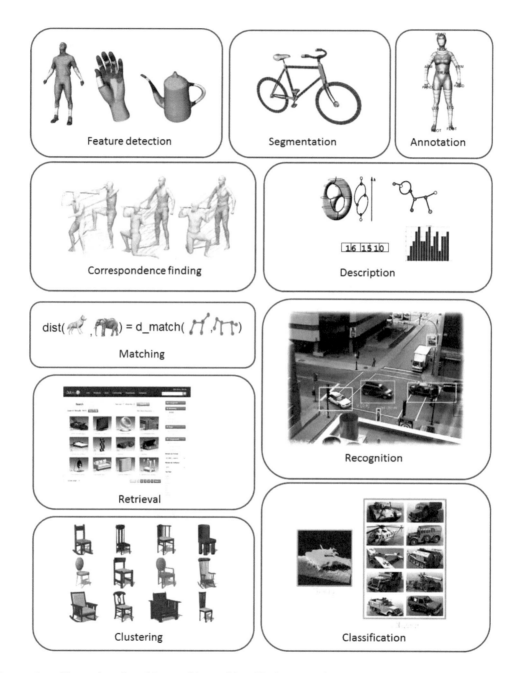

Figure 2.1: Examples of problems addressed by 3D shape analysis.

first step in the pipeline of other shape analysis solutions, such as shape description and retrieval. For example, in chapters 4 and 8 we will see how differential geometry provides tools for the identification of significant morphological features: curvature can be used to characterize surface features such as tips, pits, mounts, and blends, whereas the critical points of the height function can be used in digital terrain modeling to identify salient features such as peaks and valleys.

- **Segmentation: How can we decompose a 3D model into its significant parts?** Shape segmentation serves to decompose an object into parts which are perceptually relevant, or which have a particular meaning in a given context, for example, the base, shaft and capital in a column in 3D cultural heritage applications. Again, geometric modeling can provide tools to derive meaningful shape segmentations. For example, in chapter 5 we will see how spectral analysis gives the basis to define a library of deformation-invariant and multi-scale shape segmentations which are well-aligned with perceptual properties and intuition.

- **Semantic labeling (annotation): How can we assign meaningful textual information to objects and their parts?** Semantic annotation is the automatic, or semi-automatic, labeling of objects (or parts of objects) with a tag describing their content. It requires deriving high-level information from low-level properties. In chapters 4 and 12 we will see how the analysis of the geometry of parts, together with the study of topological relationships between shape parts (such as adjacency), gives cues to the automatic recognition and labeling of functional parts of man-made 3D shapes, for example, the tagging of vase parts with labels such as container, handle, base, tip.

- **Registration (alignment, correspondence finding): How can we align features of 3D models?** Finding correspondences between discrete sets of points on different models has applications in object tracking, surface completion, statistical shape modeling, symmetry analysis, shape interpolation, attribute transfer. It can also be part of a shape matching pipeline. In chapter 6, we will show how the study of transformations between mathematical spaces (isometries and Möebius transformations) will give rise to an efficient algorithm for discovering dense sets of point correspondences between deformable objects.

- **Description: How can we communicate what a 3D object looks like, or what is its meaning?** Shape descriptions aim to find concise yet informative *signatures* of shape models. These signatures are machine-understandable indexes to the informative content of 3D models. They enable shape matching, retrieval, and classification. For example, in chapter 5, it will be shown how a concise shape signature can be derived from the solution of the heat equation, which represents the amount of heat transferred from a point to another point in a given amount of time. The simulation of physical processes on surfaces and the use of spectral analysis techniques produces a signature which is multi-scale, deformation invariant, and robust to noise.

- **Matching (comparison): How can we compute a measure of similarity between 3D models?** The evaluation of shape similarity finds application in several problems, from product design to computational biology. Usually, shape comparison is carried out by first abstracting 3D objects as shape descriptors (or signatures), then using shape descriptors to perform similarity assessment. In chapter 7, algebraic topology helps casting the complex problem of comparing two shapes as the comparison of combinatorial objects, which turns out to be computationally efficient and stable to noise, as demonstrated by applications such as the monitoring of the effects of orthodontics treatments.

- **Retrieval: How can we find a given 3D model in a large database?** The increasing availability of powerful modeling software and 3D acquisition devices has led to rapidly growing repositories of 3D models. For example, the Trimble 3D Warehouse[1] contains millions of 3D models in many different classes and thus should be a valuable resource for people working with 3D objects. Yet, the task of exploring such large 3D repositories and retrieving the models of interest remains an important and challenging problem. One key challenge is that users may be interested in exploring different types of shape properties. The definition of shape descriptors coding different attributes, along with proper comparison strategies, is the first step towards the definition of effective 3D retrieval techniques. In chapter 12 we will see how the combination of tools from topological persistence and differential geometry yielded to a technique for 3D textured shape retrieval, which is able to analyze both colorimetric and geometric properties.

- **Classification and clustering: How can we assign a 3D model to its proper class?** Classifying and clustering 3D shapes is an important issue for many applications. For example, in 3D cultural heritage the shape-based categorization of unknown objects, such as vessels or coins, helps their attribution to given time periods, to a given society, or to a given author. In chapter 9 an example of multi-class 3D classification technique is presented, which is based on the description of 3D objects via spectral features computed on topological skeletons and augmented by a feature selection technique.

2.2   APPLICATIONS

The entertainment industry has been the driving force behind the explosion of applications and technological advances in 3D media: highly efficient GPUs are now available at a reasonable price, and commodity hardware is connected to the internet at high speed. The technological infrastructure is now ready to allow the sharing, creation and storage of huge volumes of 3D data over the web. The creation of 3D models is within reach of everybody and not only of experienced professionals as it was before, and sources for really heavy amounts of 3D data like scanning, photogrammetry or procedural/parametric design make it even easier to produce digital shapes.

[1]https://3dwarehouse.sketchup.com/

We may say that the modeling bottleneck is overcome, or at least, we are moving fast in this direction. Large collections of 3D models can be readily built up for different applications and not only for fancy-looking graphics used in entertainment. Indeed, this technological revolution has caused a gradual paradigm shift in various applied and scientific fields: from physical prototypes and production to virtual prototypes and simulation. This shift has had an enormous impact in domains where 3D media are essential knowledge carriers and represent a huge economic factor. Such domains include design and manufacturing, gaming and simulation, cultural heritage and archaeology, medicine and bioinformatics, geographic information systems and environment. Here we list a set of them, where the workflow for 3D content creation is becoming completely digital and 3D shape analysis methods are required, possibly in conjunction with other technologies, to solve real-world problems. This list, though not exhaustive, should give the reader the feeling of how diverse the applications in the scope of 3D shape analysis are.

Examples of application areas

- **Product modeling:** it is the application sector that mostly contributed to the development of techniques for modeling and processing digital 3D models. It can be informally defined as the whole workflow that stretches from an idea about a new product (e.g., an appliance or a car), to the concept development and shape design, and then to a series of engineering-related steps such as testing, manufacturing or machining the physical object. Recently, many software tools incorporated shape analysis techniques for supporting the modeling pipeline. For example, they started including shape-based retrieval techniques to reduce the human effort (and the associated costs) to look for similar parts in catalogs, independently of the Product Life Management system.

- **Archaeology and cultural heritage:** while the potential of 3D digital content in cultural heritage is clear by now, the practical exploitation has been starting quite recently. In fact, the academy first and museums after have started to adopt digital information both for the e-documentation of the past in 3D and for an effective organization and presentation of archaeological/cultural heritage content to virtual visitors. 3D shape analysis is very important as it allows the implementation of several applications such as virtual reconstruction of heritage sites; fast fragment assembly; archiving and retrieval; web-based cultural heritage; virtual study and restoration of ancient artifacts, which can be shared by different experts and shown to the public, while preserving the real object from further damages.

- **Medicine:** the recent progress in the acquisition of medical images allows one to generate 3D data that is more and more accurate and reliable. Many clinical applications such as radiotherapy planning, image-guided surgery and application of implants heavily rely on the analysis and processing of 3D content. In this context, the current trend is that of substituting generic anatomical atlases with digital patient-specific data which are crucial for simulating and monitoring surgery and therapy according to all possible anatomical variations.

Hence the medical field is facing the generation of increasingly complex and heterogeneous 3D content, which requires the help of new technologies for being classified, stored and retrieved efficiently. For example, morphometric analysis of shape data, acquired in several modalities, is suggested as a key process for the diagnosis of healthy or diseased patients.

- **Biology and bioinformatics:** shape models play a fundamental role in the scientific processes related to the understanding of bio-systems. For example, the possibility to visualize, analyze and compare molecular surfaces in a computational environment highly improves the capabilities of the scientists and offers new insights into very complex problems. For example, the geometric shape of molecular surfaces strongly influences the docking processes where shape complementarity is a necessary condition to define the binding affinity of molecules.

- **Gaming:** the great majority of real-time games, no matter the genre, take place in a 3D virtual world, and every game entity (player controlled avatars, non-player controlled characters or objects) have a position in this world, where a high level of realism is often essential to beat competitor titles, to offer a higher sense of immersion or maximize the learning experience. The computer game industry is constantly growing and game sales have already outstripped cinema box office revenues. Also serious games, designed for a primary purpose other than pure entertainment, are gaining increasing importance in defense, education, scientific exploration, health care, emergency management, city planning, engineering, religion, and politics. Simulations games are conceived as mathematical/algorithmic models which allow prediction and visualization of the evolving conditions over time on the base of the players' decisions and actions. In simulation learners typically explore a model of a particular process or phenomenon with the goal of developing certain skills without forcing users to travel through time or space or face physical risks. A common aspect of these categories of games is that they make a massive use of 3D shape data, and 3D shape analysis is fundamental to improve the gaming experience.

- **Geographic systems and environment:** the technologies that physical geographers use in their efforts to learn more about Earth were only a dream 30 years ago. The amounts of data, information, and imagery available for studying Earth and its environments have exploded. Graphic displays of environmental data and information and increased computer power allow the presentation of high-resolution images, three-dimensional scenes, and animated images of Earth features, changes, and processes. Technology may provide maps, images, and data, but tools which extract geographical aspects of the subject being studied are essential to solve the problems of understanding our planet and its environments. Efficient and effective approaches for the analysis and understanding of large-scale 3D environments represented by heterogeneous data (radar, satellite, GPS) are needed.

CHAPTER 3

Geometry, Topology, and Shape Representation

In this chapter, we introduce a number of concepts, focusing on those which are common to most of the following chapters. Many more concepts will be introduced later on, to support the formalization of advanced techniques for shape analysis.

We begin with the definition of *metric spaces*, that is, spaces where a function, a *metric*, is defined to formalize the notion of distance between points. The intuition we have of distance is rooted in the 3D Euclidean space we live in, where the distance between two points is the length of the straight line that connects them. Yet there are other ways to measure distances: imagine you are given two points on the boundary of a 3D object and you are asked to measure the length of the shortest path one would follow if bounded to walk on the boundary of the object itself. This way to measure distances yields, in general, a different result than measuring along a straight line, and it generalizes the Euclidean distance to *geodesic distances* in *curved* spaces.

Topological space is another basic mathematical concept which serves to model shapes and generalizes to arbitrary spaces concepts we are familiar with, as they are proper to the Euclidean space we live in. These concepts include, for example, those of *closeness, connectedness, continuity*. Building on topological spaces, a term we will encounter several times throughout the book is *manifold*: manifolds are the mathematical expression of spaces with a well-behaved structure and smoothness degree.

Continuity is among the first concepts we learn in basic mathematical courses and is an example of well-behaved mapping between a domain and its co-domain. This notion may be extended further to define the theoretical framework within which two spaces, or shapes, might be considered equivalent. *Functions between topological spaces* of a certain type allow us to consider two shapes as if they were the same: this is a powerful technique as we may want to perform the analysis not in its original shape space but in some transformed space, where computations or reasoning might be simpler.

3.1 METRIC AND METRIC SPACES

We are going to start our mathematical journey introducing the notion of *metric*, which has to do with a basic idea in human experience, namely, the idea of distance. In everyday life, the term *distance* means some degree of closeness of two physical objects or concepts (e.g., in space or time), and the term *metric* usually stands for a measurement. The mathematical meaning of this term originated a century ago in the work by M. Frèchet (1906) and F. Hausdorff (1914), who introduced the theory of *metric spaces*. Readers interested in more details on notations and definitions presented in this section are referred to the book [61].

Let X be a set. A function $d : X \to X$ is called a **metric** on X if, for all $x, y, z \in X$, the following holds:

1. $d(x, y) \geq 0$ (*non-negativity*)

2. $d(x, y) = 0$ if and only if $x = y$ (*identity*)

3. $d(x, y) = d(y, x)$ (*symmetry*)

4. $d(x, y) \leq d(x, z) + d(z, y)$ (*triangle inequality*)

A **metric space** (X, d) is a set X equipped with a metric d.

The Euclidean 3D world we live in is an example of a metric space, where the metric is given by the well-known Euclidean distance, that is, the distance between two points is the length of the straight line that joins them. In Figure 3.1, on the left side, the dashed segment measures the Euclidean distance between the yellow points. But imagine the shape is embedded in the empty space and the distance has to be measured by walking along the shape boundary: the distance would change considerably (see the dashed line on the right of the figure).

Figure 3.1: On the left side, the red segment between the yellow points measures the Euclidean distance, while on the right side it measures the distance on the object boundary.

We will now introduce formally the mathematical notion of *geodesic distance*, which generalizes the notion of distances in a flatland to distances in curved spaces. The term *geodesic* comes indeed from the science of measuring the Earth, namely geodesy, and stems from the fact that measures on the curved surface of the Earth require some more considerations.

3.2 GEODESIC DISTANCE

To define the geodesic distance, we need another intuitive geometric notion: *curve*. In mathematical terms, curves are continuous functions, which map an interval of the real numbers onto a subset of the target space, that is, the space where measures are necessary. The intuition behind this is that we can travel on the curve by moving on a one-dimensional interval of the real numbers. On a curved space, without the straightness constraint, there are many paths to travel from one point to another. In order to quantify the distance between two points on curved spaces, it is necessary to introduce the concept of shortest (curved) path which will be the one along which the measure will be computed. This is stated formally as follows:

Curve   Let (X, d) be a metric space. A *curve* $\gamma : I \rightarrow X$ is a continuous function defined on an interval $I \subset \mathbb{R}$. γ is called *regular* of class C^r if it is r-times continuously differentiable, and it is called *smooth* if $r = \infty$. A curve is a *path* if it does not cross itself, that is, γ is injective.

For a regular curve $\gamma : [a, b] \rightarrow X$ let us define the parameter $s = s(t) = l(\gamma|_{[a,t]})$, with $l(\gamma|_{[a,t]})$ the length of the curve portion corresponding to $[a, t]$. A curve $\gamma = \gamma(s)$ parametrized according to s is called *parametrized by arc length*, and $s = s(t)$ is also known as the curvilinear abscissa of the curve (see Fig. 3.2).

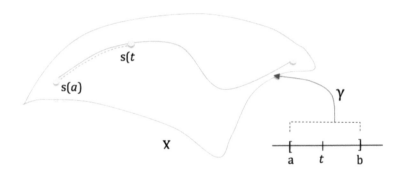

Figure 3.2: A curve γ over a surface X and its parametrization by arc length.

Curve length   The *length* $l(\gamma)$ of a curve $\gamma : [a, b] \rightarrow X$ is defined as $sup \sum_{i=1}^{n} d(\gamma(t_{i-1}, \gamma(t_i)))$ with the supremum taken over all finite decompositions $a = t_0 < t_1 < \ldots < t_n = b$, $n \in \mathbb{N}$, of $[a, b]$. A curve with finite length is called *rectifiable*.

The expression above for the curve length states that we can measure the length of the curve using the distance defined on the metric space X; we have to *cut* the interval $[a, b]$ into sub-intervals, and measure the distance of these sub-interval extrema under the mapping of γ on X. Therefore, the length of a curve $\gamma : [a, b] \rightarrow X$ is at least the distance between its end points, that is, $l(\gamma) \geq d(\gamma(a), \gamma(b))$, because the definition above mentions a subdivision process and a supremum selection over all decompositions, while the value $l(\gamma) = d(\gamma(a), \gamma(b))$ can be obtained with the trivial decomposition $t_0 = a, t_1 = b$. In case the curve is expressed by its arc length parametrization, the curve length takes the form of an integral measure.

Shortest path   Now, is there a way to move from one point on X to another one by moving along a path as short as possible? The answer is yes, and the way to do it is to follow the so-called *shortest path* between the two points. The definition is the following:

The curve γ which satisfies $l(\gamma) = d(\gamma(a), \gamma(b))$ is called the *shortest path* between $\gamma(a)$ and $\gamma(b)$.

Not all paths between two points are shortest paths: in Figure 3.3, left side, we see two paths over a sphere between two points P and Q but only the blue one represents a geodesic segment. Note that there may not be a shortest path, for instance when the shape is not connected by arcs. When existing, the shortest path may not be necessarily unique: in Figure 3.3, the picture on the right side shows that there are two shortest paths between P and Q as the distances to travel around the Gaussian bell are the same.

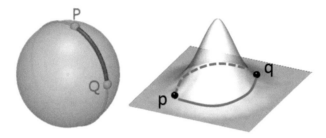

Figure 3.3:   Left: the blue line represents the geodesic segment between the points P and Q while the yellow one is a possible path whose length is not minimal. Right: the shortest path is not necessarily unique.

Geodesic distance   With these premises, the *geodesic distance* between two points is defined as the length of a shortest path between the two points. The geodesic segment and the geodesic curve are concepts which relate to quite a considerable mathematical machinery. There are subtle differences between the two, which largely depend on the structure of the space S. There are indeed metric spaces where the limit process referred to in the definition of curve length may yield unexpected results. For readers interested in additional mathematical details, we suggest the following books [35, 62].

As we will see in the following chapters, the notion of geodesic distance is quite important for shape analysis and the existence of metrics defined by the geodesic distance characterizes strongly the nature of the space where they are defined.

Intrinsic metric   Given a metric space (X, d) in which every pair of points is joined by a rectifiable curve, the *internal metric* δ on X is defined as the infimum of the lengths of all rectifiable curves connecting two given points. The metric d is called the **intrinsic metric** if it coincides with the internal metric δ. A metric space with the intrinsic metric is called an *intrinsic* or *inner metric space*.

3.3   TOPOLOGICAL SPACES

With metric spaces and distances we have the formal definitions needed to measure sizes: topology is necessary now to formalize the concept of closeness of points and to study the nature of spaces in terms of the adjacency between its points. Topological spaces are mathematical structures that generalize concepts such as closeness, limits, connectedness, or continuity, from the Euclidean space $\mathbb{R}^n$ to arbitrary sets of points. This is achieved associating to the space a formal structure of sub-sets of the space together with their relationships, rather than distances between points. To deepen the understanding of the concepts presented here, we suggest the following textbook [212].

A *topological space* (X, τ) is a set X on which a *topology* τ is defined, that is, a collection of subsets of X, which are called the *open* subsets of X and which satisfy the following axioms:

1. both X itself and the empty subset must be among the open sets: $X \in \tau$, $\emptyset \in \tau$;

2. the intersection of two open sets is open: if $A, B \in \tau$, then $A \cap B \in \tau$;

3. all unions of open sets are open: for any collection A_{ii}, if all $A_i \in \tau$, then $\cup_i A_i \in \tau$.

The complements in X of open sets are called *closed sets*.

There are two trivial examples of topology we may think of: the coarsest and the finest. The coarsest topology is the *trivial topology*, which has only two open sets, namely the empty set and X. The finest topology, the *discrete topology*, contains all subsets of X as open sets. Another example, which is intuitive for everyone, is the Euclidean topology defined by intervals. In $\mathbb{R}$, an open interval is a set $(a, b) = \{x \in \mathbb{R} | a < x < b\}$, $a \in \mathbb{R} \cup \{-\infty\}$, $a \in \mathbb{R} \cup \{+\infty\}$; the topology defined through the open intervals over $\mathbb{R}$ is called *Euclidean* topology.

Given a topological space (X, τ), a *neighborhood* of a point $x \in X$ is a set containing an open set which contains x. Open sets and distances contribute together to the characterization of spaces whose points can be separated: a topological space is called a *Hausdorff space* if and only if every two points x, y have disjoint neighborhoods.

Metric and topological spaces   The notions of metric and topological spaces are tightly connected. Any space equipped with a metric, (X, d), can be turned into a topological space (X, τ_d), with a quite simple procedure which relies on the metric d for building the structure of open sets.

Define the *open ball* of center x and radius r as the set $B(x, r) = \{y \in X : d(x, y) < r\}$, with $x \in X$ and $r \in \mathbb{R}$, $r > 0$. A subset of X which is the union of (finitely or infinitely many) balls is called an *open set*. Equivalently, a subset U of X is called an *open* set if, given a point $x \in U$, there exists a real number $\epsilon > 0$ such that, for any point $y \in X$ such that $d(x, y) < \epsilon$, $y \in U$. Now, we can define the topology τ_d as the collection of open sets defined as above, then (X, τ_d) is a topological space with the *metric topology* induced by d.

With the previous notation, the open balls of the Euclidean space $\mathbb{R}^n$ are then defined as $B(x, r) = \{y \in \mathbb{R}^n : ||x, y|| < r\}$, with $x \in \mathbb{R}^n$ and $r \in \mathbb{R}$, $r > 0$.

3.4   CONTINUOUS AND SMOOTH FUNCTIONS BETWEEN TOPOLOGICAL SPACES

The continuity of functions is the building block to reason in terms of *equivalence* between spaces. We will see in the coming chapters that equivalence between spaces is exploited many times in shape analysis: shifting the analysis from one space to another, we may simplify our reasoning thanks to theoretical results that hold in the new space. The important point is to understand under which conditions we may consider two spaces equivalent: the way to go is continuity and smoothness.

Let us recall that a function between topological spaces is said to be *continuous* if the inverse image of every open set in the co-domain space is open in the domain space. This simple and elegant formulation of continuity may be further strengthened by asking that both the function and its inverse are continuous: a *homeomorphism* is a bijection that is continuous and whose inverse is also continuous. From the viewpoint of topology, if there exists a homeomorphism between two spaces, then the spaces are essentially identical. Figure 3.4 shows examples of homeomorphic and non homeomorphic spaces.

If we want something more than topological equivalence, we have to introduce another piece of mathematics, which captures the equivalence between spaces also at differential level. The reader interested to deepen the concepts listed in the remainder of this chapter can refer to the books [80, 94, 149].

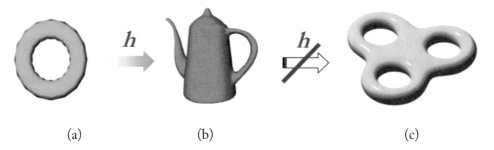

(a)   (b)   (c)

Figure 3.4: The spaces in (a) and (b) are homeomorphic while the example in (c) is not; indeed, the third model has a different number of holes and cannot be obtained from the previous ones by continuous deformations.

Let X be an arbitrary subset of $\mathbb{R}^n$. Then a function $f : X \rightarrow \mathbb{R}^m$ is called *smooth* if for every point $x \in X$ there is an open set $U \subseteq \mathbb{R}^n$ and a function $F : U \rightarrow \mathbb{R}^m$ such that $F = f_{|X}$ on $X \cap U$ and F has continuous partial derivatives of all orders.

In particular, given $X \subseteq \mathbb{R}^n$ and $Y \subseteq \mathbb{R}^m$, the smooth function $f : X \rightarrow Y$ is a *diffeomorphism* if f is bijective and $f^{-1} : Y \rightarrow X$ is also smooth. If such a function exists, the spaces X and Y are *diffeomorphic*; in this case they are intrinsically equivalent because they may be considered two copies, with two different coordinate systems, of the same abstract space.

3.5   MANIFOLDS

We introduce now another notion which is particularly relevant in shape representation and analysis: the concept of *manifold*. This term usually identifies a space "*in which the neighborhood of each point is just like a small piece of Euclidean space*" [94]. In particular, $M \subseteq \mathbb{R}^n$ is a *k-dimensional manifold* if it is locally diffeomorphic to $\mathbb{R}^k$.

Manifolds of dimension 2 and 3 are typically the mathematical models used for 3D object representation in geometric modeling. The manifold structure is used to tailor representation schemes which can ensure validity of the digital models and to define operators, such as the Euler operators, that can modify the representation while guaranteeing that at each step we still maintain the manifold structure of the representation [167].

The notions underlying the theory of manifolds builds on real analysis: we refer the interested reader to the introductory book [118] and the real analysis book [123]. For the sake of the mathematics described in the remainder of this chapter, we simply recall the definition of smooth functions, also called *analytic*, which are those belonging to the class $\mathcal{C}^\infty$.

Manifold without boundary   A topological Hausdorff space M is called a *k-dimensional topological manifold* if each point $m \in M$ admits a neighborhood $U_i \subseteq M$ homeomorphic to the open disk $D^k = \{x \in \mathbb{R}^k \mid \|x\| < 1\} \subseteq \mathbb{R}^k$ and $M = \bigcup_{i \in \mathbb{N}} U_i$.

Manifold with boundary   A topological Hausdorff space M is called a *k-dimensional topological manifold with boundary* if each point $m \in M$ admits a neighborhood $U_i \subseteq M$ homeomorphic either to the open disk D^k or the open half-space $\mathbb{R}^{k-1} \times \{x_n \in \mathbb{R} \mid x_n \geq 0\} \subseteq \mathbb{R}^k$ and $M = \bigcup_{i \in \mathbb{N}} U_i$. The number k represents the *dimension* of the manifold.

A compact[1] manifold without boundary is also called *closed*. Points on a manifold with boundary are classified either *interior*, if they have a neighborhood homeomorphic to an open disk, or *boundary*, if their neighborhood is homeomorphic to a half-disk. Figure 3.5 represents an example of a 2-manifold without boundary (a bitorus), a 2-manifold with two boundaries and a surface that in correspondence of the intersection between the plane and the bi-torus is non-manifold.

Examples of three-dimensional manifolds with boundary are the solid sphere and the solid torus; while their boundary, the usual sphere S^2 and the torus T^2, are two closed 2-manifolds. In the case of a terrain surface, which is usually modeled by a single-valued function, the reference manifold M is a two-manifold with boundary, where all points, except those along the boundary, have a neighborhood homeomorphic to a disk, see examples in Figure 3.6.

---

[1]The notion of compactness generalizes the property of the subsets of the Euclidean spaces of being closed and bounded. For a formal definition we refer to standard books of topology, e.g., [149].

Figure 3.5: From left to right: neighbors of points on a 2-manifold without boundary and a 2-manifold with boundary. In the last example, the neighbors of the points that lie in the intersection of the bitorus with a plane are not homeomophic to any disk and therefore the surface is non-manifold [217].

Figure 3.6: From left to right: a 3-manifold with boundary, a 2-manifold with boundary and a 1-manifold without boundary (a circle) [217].

3.6   CHARTS

Points on manifolds have a local neighborhood structure, which can be exploited to build a kind of local representation of the manifold itself. As curves can be parametrized by arc length, here we may construct a step-wise representation of the manifold by associating a homeomorphism $\varphi_i : U_i \to D^k$ to each open subset U_i.

Each pair (U_i, φ_i) is called a *map*, or a *chart*, while the union of charts $\{(U_i, \varphi_i\}$ is called the *atlas* on the manifold M. The terminology used here clearly reflects the meaning of these concepts: the most natural intuition we may think of are the atlases for representing the Earth.

To use charts when reasoning in the smooth domain, we need to ensure that smoothness while moving from one chart to another. For this, we need to introduce the concept of *transition function* to each atlas on a manifold. Let U_i, U_j be two arbitrary charts and $U_i \cap U_j$ be their intersection. On this intersection two coordinate maps $\varphi_i : U_i \cap U_j \to \varphi_i(U_i \cap U_j) \subset D^k$ and $\varphi_j : U_i \cap U_j \to \varphi_j(U_i \cap U_j) \subset D^k$ are defined. Since the composition of homeomorphisms is a homeomorphism, the homeomorphisms $\varphi_{i,j} : \varphi_i(U_i \cap U_j) \to \varphi_j(U_i \cap U_j)$ such that $\varphi_{i,j} = \varphi_j \cap \varphi_i^{-1}$ are well defined on the open subset $\varphi_i(U_i \cap U_j) \subset D^k$ and are called *transition functions* or *gluing functions* on a given atlas.

Figure 3.7 represents the transition functions between two charts defined on the same manifold.

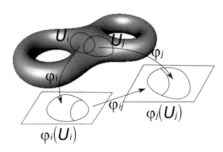

Figure 3.7: Given two neighborhoods U_i and U_j, each one with an associated coordinate map φ_i and φ_j respectively, the transition function $\varphi_{i,j}$ is defined on the intersection of U_i and U_j; in particular $\varphi_{i,j}$ must be compliant with both the φ_i and φ_j [217].

3.7   SMOOTH MANIFOLD

Smooth manifolds are manifolds whose transition functions have derivatives of all orders. Smooth surfaces, which are particularly important for shape analysis, belong to this class. Formally:

Smooth manifold   A k-dimensional topological manifold with (resp. without) boundary is called a *smooth manifold* or C^∞ with (resp. without) boundary, if all transition functions $\varphi_{i,j}$ are smooth. If all transition functions $\varphi_{i,j}$ are n times continuously differentiable, then the manifold is called a C^n manifold.

Smooth surface   A smooth 2-dimensional manifold with (resp. without) boundary is called a *smooth surface*, or simply *surface*, with (resp. without) boundary.

3.8   ORIENTABILITY

Orientability can be introduced in many ways. The intuition behind it is to be able to orient the normals at every point of our object in a consistent manner, for instance by using the well-known right-hand rule. Beside the standard Euclidean space, the concept of orientability can be extended to more general spaces, such as manifolds, where the role of normals is taken by the Jacobian.[2]

A manifold M is called *orientable* if there exists an atlas $\{(U_i, \varphi_i)\}$ on it such that the Jacobian of all transition functions $\varphi_{i,j}$ from a chart to another is positive for all intersecting pairs of regions. Manifolds that do not satisfy this property are called *non-orientable*.

---

[2]The Jacobian matrix extends the notion of gradient to multi-variate functions and represents the matrix of all the first-partial derivatives; the Jacobian is the determinant of the Jacobian matrix. For the notions of differentiability we refer the reader to standard books of real analysis [118, 123].

The *genus* of a connected, orientable surface is an integer representing the maximum number of cuttings along non-intersecting closed simple curves without rendering the resultant manifold disconnected, see also chapter 7. It is equal to the number of handles on it.

3.9   TANGENT SPACE

Proceeding with the generalization of intuitive notions from the Euclidean space to more general ones, we have to introduce the notion of tangent space. In the domain of geometry, the tangent space is defined by tangency conditions between curves on the manifold space, as follows.

Let M be a C^k manifold, $k \geq 1$, and $p \in M$. Fix a chart $\varphi : U \to \mathbb{E}^n$, where $p \in U \subseteq M$. Suppose that two curves $\gamma^1 : (-1,1) \to M$ and $\gamma^2 : (-1,1) \to M$ with $\gamma^1(0) = \gamma^2(0) = p$ are given such that $\varphi \circ \gamma^1$ and $\varphi \circ \gamma^2$ are both differentiable at 0. Then γ^1 and γ^2 are called *tangent* at 0 if the ordinary derivatives of $\varphi \circ \gamma^1$ and $\varphi \circ \gamma^2$ coincide at 0: $(\varphi \circ \gamma^1)'(0) = (\varphi \circ \gamma^2)'(0)$. If the functions $\varphi \circ \gamma^i : (1,1) \to \mathbb{E}^n$, $i = 1, 2$, are given by n real-valued component functions $(\varphi \circ \gamma^i)_1(t), \ldots, (\varphi \circ \gamma^i)_n(t)$, the condition above means that their Jacobian matrices $\left(\frac{d(\varphi \circ \gamma^i)_1(t)}{dt}, \ldots, \frac{d(\varphi \circ \gamma^i)_n(t)}{dt} \right)$ coincide at 0. This is an equivalence relation, and the equivalence class $\gamma'(0)$ of the curve γ is called a *tangent vector* of M at p. The tangent space $T_p(M)$ of M at a point p is defined as the set of all tangent vectors at p. Figure 3.8 represents the tangent plane of a point on a surface, which reflects the intuition of tangency that we have in Euclidean spaces. The dual of the tangent space is called the *cotangent* space.

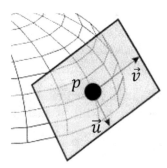

Figure 3.8: Tangent plane in p. The vectors $\vec{u}$ and $\vec{v}$ are tangent vectors [217].

3.10   RIEMANNIAN MANIFOLD

We eventually finish the list of basic notions with the Riemannian manifold, where many of the concepts introduced so far are integrated. This space, very rich in terms of geometric structure, allows us to handle properly well-behaved spaces in shape analysis and to define a set of geometric measures which have well-grounded mathematical definitions.

Let M be a real n-dimensional differentiable manifold in which the tangent space $T_p(M)$ at any point $p \in M$ is equipped with an *inner product* (i.e., a symmetric positive definite bilinear form[3] $\langle, \rangle_p$ which varies smoothly from point to point. Once a local coordinate system is chosen, every inner product $\langle, \rangle_p$ is completely defined by inner products $\langle e_i, e_j \rangle_p = g_{i,j}(p)$ of elements $e_1, e_2, \ldots, e_n$ of a standard basis in $\mathbb{R}^n$, that is, by the real symmetric and positive-definite $n \times n$ matrix $(g_{i,j}(p))$, called a *metric tensor*. In fact, if $x = (x_1, x_2, \ldots, x_n)$ and $y = (y_1, y_2, \ldots, y_n) \in T_p(M)$, we have

$$\langle x, y \rangle_p = \sum_{i,j} g_{i,j} x_i y_j.$$

The collection of these scalar products is called the *Riemannian metric g* with the metric tensor $(g_{i,j})$. The length ds of the vector $(dx_1, dx_2, \ldots, dx_n)$ is expressed by the quadratic differential form

$$ds^2 = \sum_{i,j} g_{i,j} dx_i dx_j$$

which is called the *line element* of the metric g.

A *Riemannian manifold* is a real n-dimensional differentiable manifold equipped with a Riemannian metric. The *length* of a curve γ is expressed by

$$l(\gamma) = \int_\gamma \sqrt{\sum_{i,j} g_{i,j} dx_i dx_j}.$$

The intrinsic metric on a Riemannian manifold is defined as the infimum of the lengths of rectifiable curves joining two given points of the manifold. The Riemannian metric induces the intrinsic metric of M^n, so that the distance between two points $p, q \in M^n$ is defined as

$$\inf_\gamma \int_0^1 \langle \frac{d\gamma}{dt}, \frac{d\gamma}{dt} \rangle^{\frac{1}{2}} = \inf_\gamma \int_0^1 \sqrt{\sum_{i,j} g_{i,j} \frac{dx_i}{dt} \frac{dx_j}{dt}} dt$$

where the infimum is taken over all rectifiable curves $\gamma : [0, 1] \to M^n$ connecting p and q (cf. definition of geodesic distance and inner metric before).

---

[3]A bilinear form is a function $B : X \times X \to \mathbb{R}$ which is linear in each argument separately. To every bilinear form it is possible to associate a unique quadratic form that can be expressed in terms of a square matrix.

CHAPTER 4

Differential Geometry and Shape Analysis

In this chapter, we focus on a specific type of shape: *surfaces*. Surfaces play a fundamental role in applications as they are used to model boundaries of 3D objects in computer graphics. We will draw our attention to surfaces with a geometric perspective: geometry is one of the oldest fields in mathematics, and it is concerned with questions of shape and size measurements. Geometry is still used to model objects, as our ancestors once did, but today geometric concepts have evolved to a higher level of abstraction and complexity.

What type of geometric shape properties may we compute on surfaces? The first distinction to be made is between extrinsic and intrinsic properties. *Extrinsic properties* are the properties related to how the surface is laid out in the Euclidean 3D space, and can be described by using the Euclidean distance between points; Euclidean distances form the basis for most of the earliest shape analysis methods in computer vision and computer graphics. *Intrinsic properties* are the properties related to the metric structure of the surface and are invariant to a number of deformations.

From the beginning and through the middle of the 18th century, differential geometry was studied from the extrinsic point of view: curves and surfaces were considered as lying in a Euclidean space of higher dimension (for example a surface in an ambient space of three dimensions). Starting with the work of Riemann, the intrinsic point of view was developed, in which one cannot speak of moving "outside" the geometric object because it is considered to be given in a free-standing way. Intrinsic properties started becoming more and more popular in shape analysis for their suitability to analyze deformable objects, which are ubiquitous in our reality, from human organs to living beings. Intrinsic properties can be described nicely using *geodesic distances*, which measure the length of the shortest path along the surface between two points. Geodesic and intrinsic metric were defined in section 3.2 for general spaces; here we will introduce their definition for surfaces, that is smooth 2-manifolds (section 4.1). The fundamental result concerning intrinsic properties is Gauss's *Theorema Egregium*, which elegantly connects the total curvature of a smooth manifold without boundary to the Euler characteristic of the space it represents. Another relevant result is the demonstration that the Gaussian curvature is an intrinsic invariant of surfaces and this has been used very often in applications to characterize the shape of surfaces.

4.1   GEODESIC DISTANCES ON SURFACES

In general, surfaces are defined as 2-dimensional manifolds with or without boundary. This general definition can be translated into a parametric formulation to ease the formalization of geodesic and curvature expressions.

A surface is called *regular* if in a neighborhood of each of its points it can be expressed as (*regular parametrization*)

$$\phi = \phi(u, v) = x(u, v), y(u, v), z(u, v)$$

where $\phi(u, v)$ is a regular (i.e., a sufficient number of times differentiable) vector function satisfying

$$\phi_u \times \phi_v = \begin{pmatrix} \frac{\partial x}{\partial u} \\ \frac{\partial y}{\partial u} \\ \frac{\partial z}{\partial u} \end{pmatrix} \times \begin{pmatrix} \frac{\partial x}{\partial v} \\ \frac{\partial y}{\partial v} \\ \frac{\partial z}{\partial v} \end{pmatrix} \neq 0.$$

Any regular surface can be considered as a metric space with its own *intrinsic metric*, with the *surface element* defined as

$$ds^2 = d\phi^2 = E(u, v)du^2 + 2F(u, v)dudv + G(u, v)dv^2$$

where $E(u, v) = \langle \phi_u, \phi_u \rangle$, $F(u, v) = \langle \phi_u, \phi_v \rangle$, $G(u, v) = \langle \phi_v, \phi_v \rangle$. The length of a curve γ defined on the surface by the equations $u = u(t)$, $v = v(t)$, $t \in [0, 1]$, can be computed as

$$l(\gamma) = \int_0^1 \sqrt{Eu'^2 + 2Fu'v' + Gv'^2}dt.$$

Then, the **geodesic distance** between two points on the surface is defined as the infimum of the lengths of all curves on the surface connecting the two points (cf. sections 3.1 and 3.2).

The expression

$$I = ds^2 = E(u, v)du^2 + 2F(u, v)dudv + G(u, v)dv^2$$

is called the *first fundamental form*. The *second fundamental form* is defined as

$$II = Ldu^2 + 2Mdudv + Ndv^2$$

with $L = \phi_{uu} \cdot \mathbf{n}$, $M = \phi_{uv} \cdot \mathbf{n}$, $N = \phi_{vv} \cdot \mathbf{n}$, and $\mathbf{n}$ the normal vector $\mathbf{n} = \frac{\phi_u \times \phi_v}{|\phi_u \times \phi_v|}$.

Figure 4.1 shows that the distance between two fingers measured along the hand is always the same, independently of the posture. The use of geodesic distances proved to be effective in a number of studies, and paved the road to a number of tools for intrinsic non-rigid shape analysis.

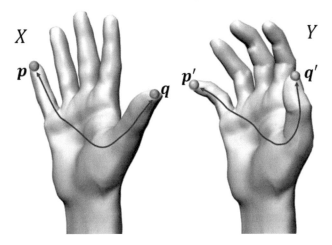

Figure 4.1: The distance measured along X between p and q is the same as the distance along Y between p' and q'.

4.1.1   COMPUTING GEODESICS ON MESHES

The computation of geodesic paths and distances on triangle meshes has been largely explored in the literature [29, 162, 191].

In Mitchell et al. [143], the authors presented an algorithm able to exactly determine the shortest path between a source and a destination on an arbitrary (possibly non-convex) polyhedral surface. The path is constrained to lie on the surface, and distances are measured according to the Euclidean metric. However, the worst-case complexity is $O(n^2 \log n)$.[1]

A popular approximation of the geodesic distance is obtained using the well-known Dijkstra algorithm [50]. In this case, the triangle mesh is modeled as a weighted directed graph whose nodes are the vertices of the mesh and the arcs correspond to the edges. Then, the shortest distance from a vertex to others is approximated by walking throughout the network starting from a source node. The shortest distance depends on the choice of the metric along the arcs (Euclidean or other cost methods). The algorithm iteratively decreases estimates on the shortest paths of non-processed vertices, which are stored in a priority queue. In each iteration of the algorithm, the closest unprocessed vertex from the source is extracted from the priority queue and processed by relaxing all its incident edges. When the algorithm starts, the length of the shortest path is overestimated and in each iteration, a shorter path is found. The algorithm ends when all nodes have been visited. The Dijkstra algorithm is computationally efficient ($O(n \log n)$) but it is constrained to pass through nodes and edges of the mesh. As a variation of the Dijkstra algorithm, Hilaga et al. [102] proposed to improve the approximation of the path by considering as edges of

---

[1]An implementation of this method is available in the MATLAB central repository, at the address: http://www.mathwork s.com/matlabcentral/fileexchange/18168-exact-geodesic-for-triangular-meshes and is also available in C language at http://code.google.com/p/geodesic/

the network also the connections between vertices that belong to two adjacent triangles and are opposite with respect to an edge.

Another popular method to solve the single source shortest path problem is the fast marching method [175, 176]. The approach proceeds by solving a discretized version of the Eikonal equation over a regular grid. The Eikonal equation is a partial differential equation measuring the first arrival time of a wavefront propagated over the grid, we refer to [5] for details. The fast marching method was extended to triangulated surfaces [115], parametric surfaces [189], and regularly sampled parametric surfaces [210].[2]

4.1.2   CONCEPTS IN ACTION

Geodesic analysis of 3D shapes   The geodesic distance is used to solve many problems of practical interest such as segmentations using geodesic balls and Voronoi regions, sampling points at regular geodesic distance or meshing a domain with geodesic Delaunay triangles; for a comprehensive review we refer to [29, 162]. Among the others, we sketch here the use of the geodesic distance to derive an intrinsic shape measure.

In general, the geodesic distance between two points can be generalized to the distance from a point p to a set of points $U \subset X$ by computing the distance from p to its closest point in X, which defines the distance map: $g_X(p) = min_{q \in X} d_g(p, q)$, where $d_g(p, q)$ denotes the geodesic distance between the points p and q. Being g_X defined as a minimum of geodesic distances, this function is neither differentiable nor smooth and therefore most of the differential geometry tools are not longer valid.

To be independent of the choice of base points, Hilaga et al. [102] introduced the average of the geodesic distance of the point from all points as an integral function. This assumption guarantees that the resulting measure is intrinsic. More formally, the value of the function f is given by:

$$f(p) = \int_{q \in X} d_g(p, q) dX$$

where q varies on X.

This function is not invariant to scaling of the object and, in case the space X is represented as a triangulation, it is replaced by its normalized representation defined as:

$$f(p) = \sum_i d_g(p, b_i) \cdot area(b_i)$$

where $b_i = b_0, \ldots, b_k$ are the base vertices for the computation of the geodesic distance which are scattered almost equally on the surface and $area(b_i)$ is the area of the neighborhood of b_i.

The resulting function is theoretically invariant with respect to rotation, translation and uniform scaling. Figure 4.2 shows the average geodesic distance for three 3D shapes. Besides

---

[2]We refer the website of the Numerical Tour Toolbox for Matlab and Scilab: https://www.ceremade.dauphine.fr/~pe yre/numerical-tour/ for examples and routines of algorithms for the computation of the shortest path.

shape comparison [102], the average geodesic distance has been used for shape decomposition and segmentation [110, 177], surface parameterization and texture mapping [214], symmetry analysis [114], partial matching [178, 197].

Figure 4.2: The behavior of the average geodesic function: the value of the function is colored from low (blue) to high (red).

4.2   CURVATURE ON SURFACES

Another property, used frequently in shape analysis, is curvature which appears in a variety of flavors in applications. For surfaces, two kinds of curvature are considered: *Gaussian curvature* and *mean curvature*. They are defined using the *principal curvatures*, which measure the maximum and minimum bending in different directions of a regular surface at each point. More precisely, at each point p of a differentiable surface in a 3-dimensional Euclidean space one may choose a unit normal vector. Each plane through p that contains the normal intersects the surface in a plane curve. This curve will in general have different curvatures for different normal planes at p.[3] The extremal values, namely the *maximal curvature k_1* and the *minimal curvature k_2*, are called the *principal curvatures* of the surface at p. A curvature is taken to be positive if the curve turns in the same direction as the surface normal, otherwise it is taken to be negative.

Figure 4.3 represents the principal curvatures k_1 and k_2 and the normal vector $\vec{n}$ at a saddle point.

The **Gaussian curvature K**, named after Carl Friedrich Gauss, is the product of the principal curvatures:

$$K = k_1 \cdot k_2.$$

The **mean curvature H** is half the sum of the principal curvatures:

$$H = \frac{1}{2}(k_1 + k_2).$$

---

[3]Recall that the curvature of a planar curve at a point p is by definition the reciprocal of the radius of the osculating circle, which is the unique circle or line which most closely approximates the curve near p.

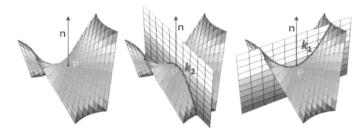

Figure 4.3: Normal vector and principal curvatures on a saddle point.

On the basis of the value of the Gaussian curvature (either positive or negative or zero) it is possible to classify a point as elliptic, hyperbolic or parabolic; see examples in Figure 4.4.

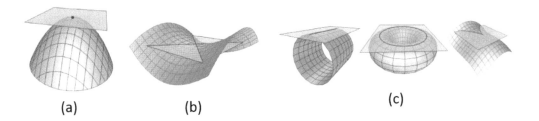

Figure 4.4: An elliptic point corresponds to $K > 0$ (a), an hyperbolic point (b) has $K < 0$, while for parabolic points (c) the Gaussian curvature is zero; if also $H = 0$ the surface is locally a plane.

The mean curvature is a purely extrinsic property, that is, it depends on how the surface is embedded in the Euclidean space $\mathbb{R}^3$. On the contrary, whereas the definition of Gaussian curvature is extrinsic in that it uses the surface's embedding in $\mathbb{R}^3$, that is normal vectors and normal planes, Gaussian curvature is in fact an intrinsic property of the surface, which does not depend on its particular embedding, but only on the Riemannian (intrinsic) metric of the surface. This surprising result is the celebrated Theorema Egregium (Latin expression meaning "remarkable theorem") which Gauss found while concerned with geographic surveys and map making: the Gaussian curvature of a smooth surface embedded in R^3 is invariant under isometric deformations of the surface (cf. Sect. 6.1). In other words, Gauss's Theorema Egregium states that Gaussian curvature of a surface can be determined from the measurements of length on the surface itself: intuitively, this means that an ant living on the surface could determine the Gaussian curvature while walking on the surface, without any reference to the external space.

The Gaussian curvature, as an intrinsic invariant, can be given entirely in terms of the first fundamental form. A simpler formula is given in terms of both the first and second fundamental forms:

$$K = \frac{LN - M^2}{EG - F^2}.$$

Finally, the integral of the Gaussian curvature over the surface region is called the **total curvature**. The total curvature is closely related to the Euler characteristic of the surface (see section 7.2), as stated in the Gauss-Bonnet theorem, which provides an important link between local geometric properties and global topological properties.

4.2.1   COMPUTING CURVATURE ON MESHES

From the computational point of view, computing curvature on triangle meshes is not trivial, because a triangular mesh is parametrized by a piecewise continuous function whose second derivatives are, almost everywhere, null. In other words, the curvature on a triangulation is concentrated along edges and at vertices, since every other point has a neighborhood homeomorphic to a planar Euclidean domain whose Gaussian curvature is null. The methods proposed in the literature for curvature evaluation can be divided into two main groups: continuity-based and property-based algorithms. The first ones transform the discrete case to the continuous one by using a local fitting of the surface which enables us to apply standard definitions. For instance, the quadratic fitting technique [137] approximates the surface with an interpolating quadratic surface and the curvatures of the quadratic surface are taken as the approximated curvatures of the triangle mesh.

The second class of algorithms defines equivalent descriptors starting from basic properties of continuous operators but directly applied to the discrete settings, guaranteeing the validity of differential properties. For example, using the *angle excess* [164] the discrete Gaussian curvature at a vertex p of a triangular mesh can be evaluated by

$$K(v) = \frac{2\pi - \sum_i \theta_i}{A}$$

where $\sum_i \theta_i$ is the sum of the angles at p in $Star(p)$ (imagine locally cutting $Star(p)$ along any of the edges incident in p, and developing $Star(p)$ onto the plane without shrinking the surface) and A is the area of $Star(p)$ or of some subregion of it containing p (see Figure 4.5). This result is consistent with the intrinsic nature of the Gaussian curvature: since this formula only takes angles into account, its value does not change under isometric deformations, that is, if the mesh is deformed preserving the length of edges (i.e., the distance between points). However, this approach is sensitive to noise and requires smoothness conditions on the input mesh.

See [86, 137, 177] and the references therein for a survey on methods for curvature estimation in the discrete setting.

4.2.2   CONCEPTS IN ACTION

Segmentation based on curvature   The automatic detection of surface features is a fundamental task in shape analysis, as it facilitates operations such as editing, matching, texturing, morphing,

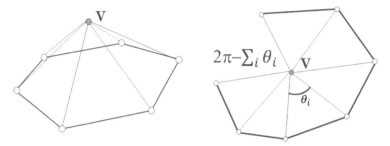

Figure 4.5: Representation of the angle excess around a vertex.

compression and simplification of 3D shapes. Several features, such as flats, tips, pits, mounts and blends, are characterized nicely in terms of curvature: for instance, pits are generally meant to be the extrema of an overall concave area, mounts have an overall convex shape without being sharp protrusions.

Going beyond the curvature alone, Mortara et al. [145] proposed a multi-scale analysis of the shape, exploiting the idea of neighborhoods of variable size, though the so-called *blowing bubbles*: a set of spheres of increasing radius, centered at each vertex of the mesh, are used to analyze the shape from a local to a more global view. The number of intersections between spheres and the shape tells how the object shape is embedded in the 3D space, while angle excess quantifies the curvature for points on protrusions of the object (see Fig. 4.6).

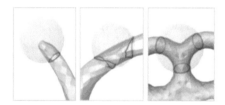

Figure 4.6: Examples of spheres intersecting shapes: the number of intersections (1, 2, 3) is related to the embedding of the shape in the 3D space.

The result is a morphological decomposition of the object (Figure 4.7), which stores features together with their morphological type, persistence at scale variation, amplitude and/or size, so that it can be used to automate processes such as shape matching and comparison.

Figure 4.7: Shape segmentation of a pot at different scales [145].

CHAPTER 5

Spectral Methods for Shape Analysis

The decomposition of a shape into simpler pieces is the ultimate goal of most mathematical methods that aim at combining synthesis with saliency. This is the peculiar characteristic of spectral methods: the decomposition and expression of a complex function into a set of simpler ones. Probably, the most popular example is Fourier analysis, which studies the way general functions may be represented or approximated by sums of simpler trigonometric functions. One best-known application of Fourier analysis is signal processing: in this case a signal is decomposed into a linear combination of the eigenvectors of the Laplace operator applied to the signal. The Laplace operator is a popular tool for modeling physical phenomena such as the diffusion equation for heat and fluid flow and wave propagation.

The first application of Fourier analysis to 3D shapes was proposed by Taubin in 1995, [193], to show how shape compression and smoothing can take advantage of the spectral analysis of the mesh geometry. This first example of mesh Laplace operator originated much further development, suggesting methods not only for analyzing but also for editing and interpolating meshes, using for instance the differential coordinates.

In general, the popularity of spectral methods derives from the flexibility and the generality of the paradigm behind Laplace operators: indeed, they project the analysis of a shape to the study of the model eigenstructures (eigenvalues, eigenvectors, or eigenspaces) derived from appropriately defined mesh operators. The recent literature is rich with successful applications of spectral analysis to graph theory, computer vision, machine learning, graph drawing, etc., and several surveys have been proposed [187, 215]; in this chapter we mainly focus on the extension of the Laplace operator and the heat equation to 3D shape analysis.

5.1 LAPLACE OPERATORS

The Laplace operator, or Laplacian, is a differential operator given by the divergence of the gradient of a function on the Euclidean space $\mathbb{E}^n$, formally:

$$\Delta f := \operatorname{div}(\operatorname{grad} f) = \nabla \cdot \nabla = \sum_i \frac{\partial^2}{\partial x_i^2},$$

where grad and div are the gradient and divergence on the space and the point $p \in \mathbb{E}^n$ is represented by the Cartesian coordinates $p = (x_1, \ldots, x_n)$. Therefore, the Laplacian requires that the function f is at least a twice-differentiable real-valued function.

Intuitively, the Laplace operator generalizes the second order derivative to higher dimensions, and it characterizes the variation of a function: $\Delta f(p)$, indeed, measures the difference between $f(p)$ and its average in a small neighborhood of $p \in \mathbb{E}^n$. In other words, the Laplacian $\Delta f(p)$ of a function f at a point p, up to a constant depending on the dimension, represents the rate at which the average value of f over spheres centerd at p, deviates from $f(p)$ as the radius of the sphere grows.

The generalization of the Laplace operator to manifolds equipped with a Riemanninan metric (see section 3.10) is called the Laplace-Beltrami operator of f and its computation requires complex calculations, which are well formalized in the *discrete exterior calculus (DEC)* [91].

The *Laplace eigenvalue problem*

$$\Delta f = -\lambda f. \tag{5.1}$$

admits as a solution an orthonormal eigensystem (ψ_i, λ_i) with non-negative eigenvalues λ_i, $\lambda_0 \leq \lambda_1 \leq \ldots \leq \lambda_i \leq \lambda_{i+1} \leq \ldots \leq +\infty$ and corresponding orthonormal eigenfunctions ψ_i.

The equation 5.1 is known in physics as the Helmholtz equation representing the spatial component of the wave equation. Thinking of a shape S as of a vibrating membrane, the eigenfunctions ψ_i can be interpreted as natural vibration modes of the membrane, while the eigenvalues λ_i assume the meaning of the corresponding vibration frequencies [198].

When the shape S is the n-sphere, the eigenfunctions of the Laplacian are the well-known spherical harmonics. More in general, the eigenbasis of the Laplace-Beltrami operator is referred to as the harmonic basis of the manifold, and the functions ψ_i as manifold harmonics [126]. The use of Laplacian eigenbases has been shown to be fruitful in many computer graphics applications and several techniques in shape analysis, synthesis, and correspondence rely on the harmonic bases that allow using classical tools from harmonic analysis on manifolds. For a detailed discussion on the main properties of the Laplace-Beltrami operator, we refer the reader to [168, 171, 209].

Several discrete Laplace operators exist [126]. When considering manifolds, the shape S is frequently approximated by a triangulation $\mathcal{T}$ with vertices $V := \{\mathbf{p}_i, i = 1, \ldots, n\}$. The function f on $\mathcal{T}$ is defined by linearly interpolating the values $f(\mathbf{p}_i)$ of f at the vertices of $\mathcal{T}$. This is done by choosing a base of piecewise-linear *hat-functions* φ_i, with value 1 at vertex $\mathbf{p}_i$ and 0 at all the other vertices. Then f is given as $f = \sum_{i=1}^{n} f(\mathbf{p}_i)\varphi_i$. Discrete Laplace-Beltrami operators are usually represented as:

$$\Delta f(\mathbf{p}_i) := \frac{1}{d_i} \sum_{j \in N(i)} w_{ij} \left[f(\mathbf{p}_i) - f(\mathbf{p}_j) \right], \tag{5.2}$$

where $N(i)$ denotes the index set of the 1-*ring* of the vertex $\mathbf{p}_i$, i.e., the indices of all neighbors connected to $\mathbf{p}_i$ by an edge. The masses d_i are associated with a vertex i and the w_{ij} are the symmetric edge weights.

In order to clarify the form assumed by 5.1 and 5.2, let us introduce some notations.

- $\mathbf{f} := (f(\mathbf{p}_1), \ldots, f(\mathbf{p}_n))^T$ is the vector of the function values at the vertices;

- $W := (w_{ij})$ the *weighted adjacency matrix* coding the vertex adjacency in the mesh;

- $V := \text{diag}(v_1, \ldots, v_n)$ the diagonal matrix whose elements are $v_i = \sum_{j \in N(i)} w_{ij}$;

- $A := V - W$ is the *stiffness matrix*;

- $D := \text{diag}(d_1, \ldots, d_n)$ the *lumped mass matrix*;

- $L := D^{-1}A$ the *Laplace matrix* (generally not symmetric).

With these notations, then the problem (5.1) can be expressed as $L\mathbf{f} = \lambda \mathbf{f}$ or better, as a generalized symmetric problem $A\mathbf{f} = \lambda D\mathbf{f}$.

Depending on the different choices of the edge weights and masses, the discrete Laplacian operators are distinguished between *geometric operators* and *finite-element operators* [169]. A deep analysis of different discretizations of the Laplace-Beltrami operator (geometric Laplacians, linear and cubic FEM operators) in terms of the correctness of their eigenfunctions with respect to the continuous case is shown in [169].

Except for some special cases (e.g., [14, 27, 100, 183]), the discrete Laplacian is not guaranteed to converge to the Laplace-Beltrami operator. In addition, when dealing with intrinsic shape properties, the Laplacian must be independent or at least minimally dependent on the triangular mesh and thus the discrete approximation must preserves the geometric properties of the Laplace-Beltrami operator. Unfortunately, Wardetzky et al. in [209] showed that for a general mesh, it is theoretically impossible to satisfy all properties of the Laplace-Beltrami operator at the same time, and thus the ideal discretization does not exist. This result also explains why there exists such a large diversity of discrete Laplacians, each having a subset of the properties that make it suitable for certain applications and unsuitable for others [35].

5.1.1 CONCEPTS IN ACTION

In the field of shape analysis and segmentation, spectral methods are extremely promising, as they naturally provide a set of tools that are intrinsically multi-scale and defined by the shape itself. Indeed, the eigenfunctions of the Laplace-Beltrami operator yield a family of real valued functions that provide interesting insights in the structure and morphology of shapes.

Shape segmentation   An example of use of the Laplace-Beltrami eigenfunctions to provide a multiscale segmentation of the shape has been proposed in [169]. There, the focus is on the *nodal sets* and *nodal domains* of the Laplace-Beltrami eigenfunctions, showing that they induce a shape decomposition which captures features at different scales, generally well-aligned with perceptually relevant shape features. The set of decompositions induced by the eigenfunctions yields the sought *library* of intrinsic shape segmentations.

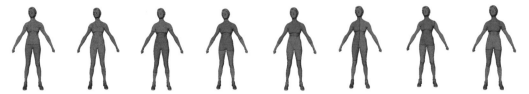

Figure 5.1: Segmentations induced by the nodal domains of some eigenfunctions selected among the first 15 eigenfunctions (in order of increasing eigenvalues). Blue regions correspond to regions where the eigenfunctions have negative values, while red regions correspond to positive values.

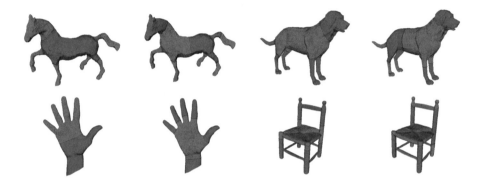

Figure 5.2: Segmentations induced by the nodal domains of some eigenfunctions selected among the first 15 eigenfunctions.

More formally, the *nodal sets* Γ_i are the zero sets of the eigenfunctions of the Laplacian operator on a Riemannian manifold, i.e., $\Gamma_i := \phi_i^{-1}(0)$. A *nodal domain* is a connected component of the complement of the nodal sets. Then, for each eigenfunction ϕ_i, the nodal sets decompose a surface into regions where ϕ_i has constant sign. In other words, each eigenfunction induces a shape segmentation, with segments corresponding to regions of positive or negative values. The use of nodal sets and nodal domains to segment 3D shapes was addressed in [125]. The first k Laplacian eigenfunctions, ordered according to increasing frequencies, provide a family of shape segmentations, each capturing different shape properties (see Figure 5.1 and Figure 5.2).

The quality of these segmentations was based on the type and correctness of the segmentation; the quality of boundaries; the definition of a multi-scale segmentation; the invariance to pose; the sensitivity to noise and tessellation; the computational complexity and the use of control parameters. Details of the discussions can be found in [169]. We simply remark that the nodal domains related to the first eigenfunctions subdivide the input surface into patches which have almost the same weight, measured as the sum of the edge weights associated with the 1-star of each vertex. In this case, the nodal sets often identify privileged directions, related to the symmetries of the objects (see also [154]). For articulated objects, the first eigenvectors define patches that are able to identify surface protrusions, and are often well aligned with perceptual features. In Figure 5.1 some segmentations of a human model induced by the nodal domains of different eigenfunctions are shown, chosen among the first 15 in the spectrum. At different scales, the segmentations capture the symmetry of the shape, the arms, legs, hands and feet of the model. Additional examples are given in Figure 5.2.

Pose transfer   The pose transfer approach proposed in Lévy [125] is based on the Fourier decomposition of the manifold embedding coordinates. There the "layout" (pose) of the shape X is transferred to the shape Y while preserving the geometric details of Y. In a formal way, given two shapes X and Y embedded in $\mathbb{R}^3$ with the corresponding harmonic bases ψ_i^X and ψ_i^Y, and the corresponding Fourier decompositions of the embeddings $X = \sum_{i \geq 1} a_i \psi_i^X$ and $Y = \sum_{i \geq 1} b_i \psi_i^Y$, a new shape Z is composed according to the rule $Z = \sum_{i=1}^{N} a_i \psi_i^Y + \sum_{i > N} b_i \psi_i^Y$ with the first N low frequency coefficients taken from X, and higher frequencies taken from Y. Figure 5.3 represents how the pose of the shape X is transferred to Y generating a new shape Y^t.

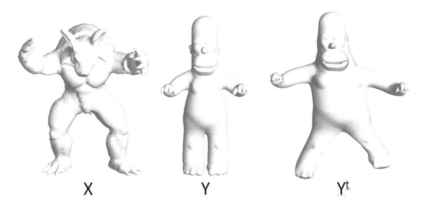

Figure 5.3: The shape Y^t is obtained from Y with the transfer of the pose of X.

5.2   HEAT EQUATION

The heat equation describes the distribution of heat (or variation in temperature) in a given region over time. In case the shape S is a compact two-dimensional Riemannian manifold, the diffusion process on S is described by the partial differential equation:

$$\left(\frac{\partial}{\partial t} + \Delta \right) f(t, x) = 0, \tag{5.3}$$

where Δ denotes the positive-semidefinite Laplace-Beltrami operator associated with the Riemannian metric of S. The heat equation governs the distribution of heat from a source point on the surface. The initial condition of the equation is some initial heat distribution $f(0, x)$ at time $t = 0$; if S has a boundary, appropriate boundary conditions must be added.

The *heat kernel* $h_t(x, y)$ is a fundamental solution of equation 5.3, with point heat source at x, and heat value at y after time t: in other words, it represents the amount of heat transferred from x to y in time t due to the diffusion process (Figure 5.4).

The value of the heat kernel $h_t(x, y)$ can also be interpreted in terms of transition probability: given a Brownian motion (continuous analog of a random walk) on S starting at point x and a subregion $C \subset S$, $\int_C h_t(x, z)dz$ is the probability that the Brownian motion will be in C at time t.

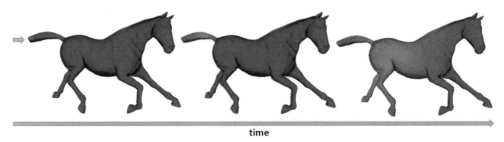

time

Figure 5.4: The heat kernel represents the amount of heat transferred from a source point in time t.

Using spectral decomposition, the heat kernel can be represented as:

$$h_t(x, y) = \sum_{i \geq 0} \exp^{-\lambda_i t} \psi_i(x)\psi_i(y) \tag{5.4}$$

Here, ψ_i and λ_i denote, respectively, the eigenfunctions and eigenvalues of the Laplace-Beltrami operator satisfying $\Delta\psi_i = \lambda_i\psi_i$.

Since the coefficients rapidly decay, the heat kernel is generally approximated by the truncated sum:

$$h_t(x, y) = \sum_{1}^{N} \exp^{-\lambda_i t} \psi_i(x)\psi_i(y). \tag{5.5}$$

The computation of the spectrum of the discrete Laplacian is the main computational bottleneck for the evaluation of the heat kernel; in fact, it takes from $O(n)$ to $O(n^3)$ operations, according to the sparsity of the Laplacian matrix. Recently, a discrete and spectrum-free computation of the diffusion kernel on a 3D shape (either represented as a triangulation or a point cloud) has been proposed in [161]. The main idea is to avoid the computation of the full spectrum adopting the Chebyshev approximation [47, 144] of the weighted heat kernel matrix. In practice, the approximation of the diffusion kernel is reduced to the solution of the sparse linear systems and a sequence of matrix-vector multiplications, without computing the Laplacian spectrum.

5.2.1   CONCEPTS IN ACTION

The heat kernel signature (HKS)   Sun et al. [190] proposed using the diagonal of the heat kernel, restricted to the temporal domain, as a local descriptor, referred to as the Heat Kernel Signature

(HKS). More precisely, given a point x on a manifold M they define its heat kernel signature $HKS(x)$ to be a function over the temporal domain:

$$HKS(x) : \mathbb{R}^+ \to \mathbb{R}, \quad HKS(x,t) = h_t(x,x).$$

In practice, they sample HKS uniformly over the logarithmic scaled temporal domain and obtain an n-dimensional descriptor vector to represent the HKS for each point:

$$p(x) = c(x)(h_{t_1}(x,x), \ldots, h_{t_n}(x,x))$$

where $c(x)$ is a scaling factor.

Figure 5.5: A color-coded (red for higher values, blue for lower values) visualization of the HKS function of isometric transformations of the same shape (left) and of a model having different topology (glued fingers, right).

The HKS descriptor has many advantages, which make it a favorable choice for shape description and matching. First, since the heat kernel is intrinsic (i.e., expressible solely in terms of the Riemannian structure of M) it is invariant under isometric deformations of the manifold (Figure 5.5, left). Second, such a descriptor captures information about the neighborhood of a point x on the shape at a scale defined by t: it captures differential information in a small neighborhood of x for small t, and global information about the shape for large values of t. Thus, the n-dimensional feature descriptor vector $p(x)$ can be seen as analogous to the multi-scale feature descriptors used in the computer vision community. Third, for small scales t, the HKS descriptor takes into account local information, which makes topological noise have only local effect (Figure 5.5, right). Note that this is a main difference with respect to the behavior of the integral geodesic distance discussed before.

Sun et al. [190] proposed different applications of the HKS. First, the local maxima of the function $k_t(x,x)$ for a large t can be used to find salient feature points. Then, by computing the L_2-norm of the difference between HKS vectors at feature points, it is possible to perform shape correspondence between different models. Finally, the HKS can be used for multi-scale self-matching: for a given point on a model, other points on the same model having HKS values within a threshold identify repeated structure on the object, possibly at different scales.

CHAPTER 6

Maps and Distances between Spaces

This chapter deals with shape *transformations*: a transformation, or *map*, is any function ϕ mapping a set X to another set Y (or to the set X itself). The simplest examples are Euclidean transformations: rotation, translation, scaling. A more elaborate question concerns the effect of maps on shape properties: indeed, transformations can be categorized according to the properties they *preserve* while moving from one space to the other. For example, *isometries* preserve the space metric structure, meaning that they preserve the distance computed between points: the distance between two points on the input space X is equal to the distance between the point images in Y (section 6.1.1). Analogously, *affinities* are transformations that preserve straight lines (section 6.1.2), and *Möbius transformations* (section 6.1.3) preserve angles.

Strictly related to shape transformations is the concept of *distances* between shapes. Indeed, one of the cornerstone problems in shape analysis is how to define a notion of shape (dis)similarity. Computing distances between shapes has fundamental applications in shape matching, recognition and retrieval. Well known examples of distances are the *Hausdorff distance* (section 6.2.1) and the *bottleneck distance* (section 6.2.2), which measure how far two subsets of a metric space are from each other; the *Gromov-Hausdorff distance* (section 6.2.3), which casts the comparison of two metric spaces as a problem of comparing pairwise distances on the spaces. Finally, the *natural pseudo-distance* (section 6.2.4) is a measure of how much shape properties are preserved while transforming a shape into another. Indeed, we may want to analyze to what extent two spaces represent two instances of some common class, up to a certain notion of invariance. Are an upright and a downright arrow instances of a common class? Are a standing woman and a sitting one similar, though their pose is different? The answer depends on the application, and on the properties we want to preserve.

6.1 SPACE TRANSFORMATIONS

6.1.1 ISOMETRIES

An isometry between metric spaces is a distance-preserving transformation: the distance between points in the image metric space equals the distance between points in the original metric space. Formally, let (X, d_X) and (Y, d_Y) be two metric spaces. A map $\phi : X \rightarrow Y$ is called an *isometry* if, for any $x, y \in X$, it holds:

$$d_Y\left(\phi(x), \phi(y)\right) = d_X(x, y). \tag{6.1}$$

The metric spaces X and Y are called *isometric* if there is a bijective isometry from X to Y. A property of metric spaces which is invariant with respect to isometries is called a *metric property* (or *metric invariant*).

In the Euclidean plane $\mathbb{R}^2$ and the Euclidean space $\mathbb{R}^3$, an example of isometries are geometric congruences, which include rigid motions (translations, rotations) and reflections. If d_X is the intrinsic metric of X, isometries preserve geodesic distances. In Figure 6.1, the hand bending is the result of an isometric transformation ϕ, and the geodesic distance between points p and q on the hand on the left is the same as the geodesic distance of their images $p\prime$ and $q\prime$ on the hand on the right, after bending.

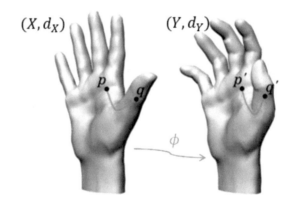

Figure 6.1: Isometric transformation of the space X into Y.

6.1.2   AFFINE TRANSFORMATIONS

An *affine transformation* (also called *affine map* or *affinity*) is a transformation ϕ between affine spaces[1] which preserves:

- the collinearity between points: points which lie on a straight line (called collinear points) before the transformation, still lie on a straight line after transformation;

- the ratio of distances between collinear points: for distinct collinear points p_1, p_2, p_3, the ratio of $\overrightarrow{p_1 p_2}$ and $\overrightarrow{p_2 p_3}$ is the same as that of $\overrightarrow{f(p_1)f(p_2)}$ and $\overrightarrow{f(p_2)f(p_3)}$;

- the barycenters of weighted collections of points; for example, the midpoint of a line segment remains the midpoint after transformation.

[1]Informally, affine spaces are a generalization of vector spaces, where there is no a single point that serves as an origin.

Affinities do not preserve, in general, angles and lengths, though they have the property that parallel lines remain parallel after transformation.

Examples of affine transformations include translation, geometric contraction, expansion, homothety, reflection, rotation, shear mapping, similarity transformation, spiral similarities and compositions of them. Some examples are shown in Figure 6.2.

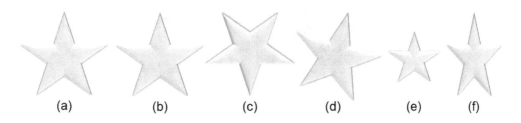

| (a) | (b) | (c) | (d) | (e) | (f) |

Figure 6.2: Some examples of affine transformations of an object (a), from left to right: congruence (identity) (b), reflection (c), rotation (d), homothety (e), and shear (f) [217].

An alternative definition of affinities involving linear algebra is based on affine combinations of points. In affine spaces, affine combinations of points are defined as linear combinations in which the sum of the coefficients is 1. Let $\{a_i\}_{i \in I}$ be a family of points in an affine space X, and $\{\lambda_i\}_{i \in I}$ be coefficients such that $\sum_{i \in I} \lambda_i = 1$. Then, $\phi : X \to Y$ is an affine map if and only if it holds:

$$f\left(\sum_{i \in I} \lambda_i a_i\right) = \sum_{i \in I} \lambda_i \phi(a_i) \,.$$

6.1.3   MÖBIUS TRANSFORMATION

A Möbius transformation is any map of the form:

$$\phi(z) = \frac{az + b}{cz + d}$$

where z is a complex variable and a, b, c, d are complex numbers satisfying $ad - bc \neq 0$.

The main properties of Möbius transformations (also called fractional linear transformations, projective linear transformations, homographies, homographic transformations, linear fractional transformations, bilinear transformations) are that they are *conformal*, that is, they preserve angles; they map every straight line to a line or circle; and they map every circle to a line or circle (see Figure 6.3).

The set of all Möbius transformations forms a group under composition called the Möbius group.

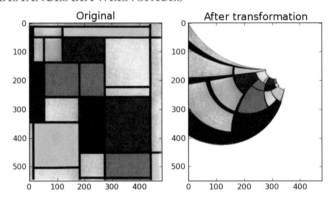

Figure 6.3: A Möbius transformation of an image (based on `http://i.imgur.com/qz7P0.png`).

6.1.4   CONCEPTS IN ACTION

3D shape correspondence and symmetry detection via Möbius maps   Finding correspondences between a discrete set of points on two surface meshes is a fundamental problem of 3D shape analysis, with applications in shape interpolation, attribute transfer, surface completion, statistical shape modeling, symmetry analysis, shape matching, and deformable surface tracking. For many of these applications, the input meshes represent different objects in different poses: they are often approximately isometric, or are composed of large parts that are nearly isometric. In this case, the problem is that of finding corresponding points such that the mapping between them is close to an isometry (cf. section 6.1).

The authors of [128] developed an efficient algorithm for discovering dense sets of point correspondences between surfaces that are approximately and/or partially isometric. The key observation is that isometries are a subset of the Möbius group, whose properties make computationally tractable the problem of looking for a correspondence map g between two genus zero surfaces[2] (Figure 6.4).

Indeed, any genus zero surface can be mapped conformally (with angles preserved) to the unit sphere (maps ϕ_1 and ϕ_2 in Figure 6.4). Therefore, any isometry g between genus zero surfaces gives rise to a one-to-one and onto conformal map from the unit sphere to itself (the composition of the inverse of ϕ_1, g and ϕ_2 in Figure 6.4).

One-to-one and onto mappings of a sphere to itself are Möbius transformations. The Möbius group is known to have six degrees of freedom, that is, fixing three distinct points on each sphere defines a Möbius map uniquely. Moreover, the Möbius transformation that interpolates any three points can be computed in closed-form. Finally, deviations from isometry can be modeled by a transportation-type distance between corresponding points.

According to these observations, [128] developed a technique for finding point correspondence between nearly isometric surfaces (Figure 6.5). The algorithm iteratively:

---

[2]Informally, a surface without donuts-like, passing holes; see chapter 3 for a formal definition.

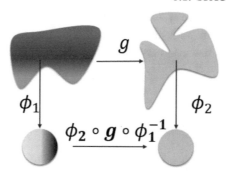

Figure 6.4: Casting the correspondence problem between two surfaces as the Möbius transformation between unit spheres.

1. Samples three points from each mesh; this is done by selecting the local maxima of Gauss curvature then adding the (geodesics) farthest points;

2. Computes the Möbius transformation that aligns the three point pairs;

3. Transforms all (sampled) points from both meshes by that Möbius transformation;

4. Evaluates the intrinsic deformation error between mapped points (deviation from isometry);

5. Produces "votes" for predicted correspondences between the mutually closest points with magnitude representing their estimated deviation from isometry.

The computed deformation errors are accumulated in a fuzzy correspondence matrix, which can be analyzed to determine a consistent set of discrete correspondences through a voting strategy.

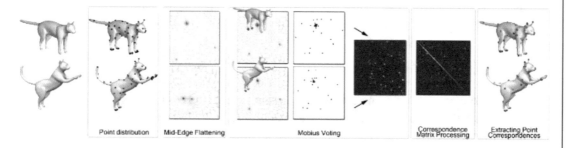

Figure 6.5: Pipeline of the shape correspondence algorithm [128].

The algorithm works in polynomial time and is (theoretically, in the smooth case) guaranteed to find the optimal set of correspondences for perfect isometries and extends well to near isometries. The experiments in [128] show that it can find intrinsic point correspondences in cases of extreme deformation (Figure 6.6). The drawbacks are that it can only deal with genus zero surfaces, and it may suffer from bad artefacts when the meshes in consideration have "bad" triangles.

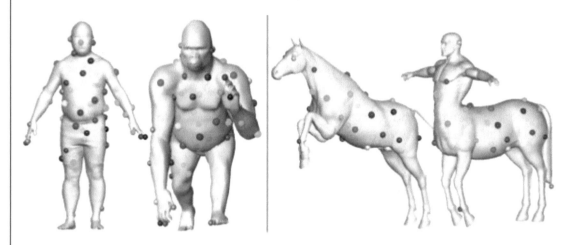

Figure 6.6: Correspondences between the models of a gorilla and a man (left), and of a centaur and a horse.

Similar ideas led in [114] to an algorithm for detecting intrinsic symmetries on a surface mesh, a problem which has applications in surface compression, completion, matching, beautification, alignment. Indeed, looking for intrinsic symmetry transformations means finding isometric transformations that map a surface onto itself (self-isometries), which are contained in a group of Möbius transformations. Based on these observations, the algorithm 1) generates a set of symmetric points by detecting critical points of the Average Geodesic Function, 2) enumerates small subsets of those feature points to generate candidate Möbius transformations, and 3) selects among those candidate Möbius transformations the one(s) that best map the surface onto itself.

6.2   DISTANCES BETWEEN SPACES

From the problem of comparing subsets of the same metric space (sections 6.2.1 and 6.2.2) we move towards the comparison of two metric spaces (i.e., two spaces possibly equipped with different distances, section 6.2.3) or two topological spaces, section 6.2.4. First we start to show how in a metric space (M, d), distances between points can originate distances between subsets. A first example is the Hausdorff distance, a further example is the *bottleneck* distance, also known as *matching* or *multiset-multiset* distance.

6.2.1   HAUSDORFF METRIC

The Hausdorff distance between two subsets of a metric space M is the greatest of all the distances from a point in one subset to the closest point in the other subset. In other words, two subsets are close in the Hausdorff distance if every point of either subset is close to some point of the other subset. Formally, the Hausdorff metric (also known as *two-sided* Hausdorff distance) $d_H(X, Y)$ between two non-empty subsets X and Y of the metric space (M, d) is defined as:

$$d_H(X, Y) = \max\{ \sup_{x \in X} \inf_{y \in Y} d(x, y), \sup_{y \in Y} \inf_{x \in X} d(x, y) \}, \tag{6.2}$$

where sup represents the supremum and inf the infimum. Due to its relatively easy evaluation, the Hausdorff distance is very popular for the shape comparison, ranging from images to digital terrain surfaces to 3D objects. An extension of this concept is provided by the L_p-Hausdorff distance [8] of which the Hausdorff metric represents the particular case $p = \infty$.

6.2.2   BOTTLENECK DISTANCE

The *bottleneck* or *matching* distance between two subsets X, and Y of a metric space M is defined by:

$$d_m(X, Y) = \inf_{\phi} \max_{x \in X} d(x, \phi(x)), \tag{6.3}$$

where ϕ runs over all bijections between X and Y, considered as multisets. A method to compute the matching distance has been proposed in d'Amico et al. [53]. Such a measure is useful to compare sets of points embedded in the metric space; in particular, it is used to compare and derive stability of persistence diagrams and spaces (see sections 11.2 and 11.3).

6.2.3   GROMOV-HAUSDORFF MEASURE

The Gromov-Hausdorff distance casts the comparison of two metric spaces (X, d_X) and (Y, d_Y) as a problem of comparing pairwise distances on the spaces (Figure 6.7). Notice that in this case, both the spaces and the distances might differ, e.g., d_X could be the Euclidean metric and d_Y the Riemannian metric, etc. Equivalently, the computation of the Gromov-Hausdorff distance between spaces can be posed as measuring the distortion caused by embedding one metric space into another, that is, evaluating how much the metric structure is preserved while mapping a shape into the other. By considering different metrics between points, we get different notions of metrics between spaces.

Formally, the Gromov-Hausdorff distance [92], $d_{GH}(X, Y)$, between two metric spaces (X, d_X) and (Y, d_Y) is defined as

$$d_{GH}(X, Y) = \min_{\phi_{XY}, \phi_{YX}} \max(\delta(\phi_{XY}), \delta(\phi_{YX}), \delta(\phi_{XY}, \phi_{YX})) \tag{6.4}$$

where the distortion $\delta(\phi_{XY})$ is measured using the maximum distortion induced by the map ϕ_{XY}:

$$\delta(\phi_{XY}) = \max |d_X(x, y) - d_Y(\phi_{XY}(x), \phi_{XY}(y))|$$

$x, y \in X$ (analogous definition for the distortion of the map ϕ_{YX}) and the joint distortion $\delta(\phi_{XY}, \phi_{YX})$ is:

$$\delta(\phi_{XY}, \phi_{YX}) = \max_{x \in X, y \in Y} |d_X(x, \phi_{YX}(y)) - d_Y(\phi_{XY}(x), y)|.$$

d_{GH} is a metric and satisfies the triangle inequality. This Gromov-Hausdorff distance was introduced in Computer Vision by Mámoli and Sapiro [140]. For discretized spaces X_N and Y_N sampled with the same number N of points, if one restricts its attention to bijective mappings $\phi : X_N \to Y_N$, one can approximate the Gromov-Hausdorff distance (6.4) by a permutation distance:

$$d_{GH}(X_N, Y_N) = \min_{\sigma \in \mathcal{P}(N)} \max_{0 \leq i, j \leq N} |d_{X_N}(x_i, x_j) - d_{Y_N}(x_{\sigma_i}, x_{\sigma_j})|, \tag{6.5}$$

where $\mathcal{P}(N)$ is the set of permutation of N numbers. This distance approximates equation 6.4 for randomized samplings, see [140]. However, also computing the distance 6.5 is computationally prohibitive, since it requires to check all possible permutations and an approximate algorithm was developed in [140]. Variations of the Gromov-Haussdorf distance are the L_p Gromov-Hausdorff distances and the Gromov-Wasserstein distances [138, 139].

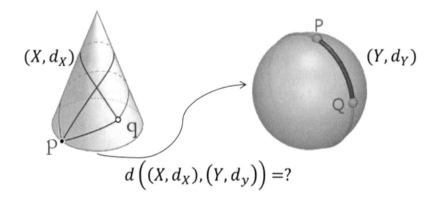

Figure 6.7: Two metric spaces X and Y can be equipped with different distances d_X and d_Y, respectively; the comparison of the couples (X, d_X) and (Y, d_Y) should measure the effort to distort one space and its metric into the other [217].

6.2.4   NATURAL PSEUDO-DISTANCE

The definition of the natural pseudo-distance pushes further the idea of measuring the distortion of properties while transforming a shape into another, which we have seen with the Gromov-Hausdorff distance.

Let us assume that a shape is a space endowed with a real function, which describes some shape properties. To compare two shapes, we can imagine transforming one shape into the other, and check how much the properties of the original shape have been preserved/distorted; this amounts to measuring how much the values of the real function representing those properties have been altered. The natural pseudo-distance offers a framework in which we can plug in different properties, in the form of different real functions, so as to measure shape (dis)similarity up to different notions of invariance.

Let us assume a shape is conceptualized as a pair (S, f) with S a topological space and $f : S \to \mathbb{R}$ a real function that measures some properties. Let (X, f_X) and (Y, f_Y) be two shapes to be compared. Formally, d_{np} is defined by setting

$$d_{np}(X, f_X), (Y, f_Y)) = \inf_{h \in H_{X,Y}} \sup_{P \in X} |f_X(P) - f_Y(h(P))|,$$

where h varies in a subset $H_{X,Y}$ of the set H of all homeomorphisms between X and Y. The subset $H_{X,Y}$ must satisfy the following axioms: the identity map $id_X \in H_{X,X}$; if $h \in H_{X,Y}$ then the inverse $h^{-1} \in H_{Y,X}$; if $h_1 \in H_{X,Y}$ and $h_2 \in H_{Y,Z}$ then the composition $h_2 \circ h_1 \in H_{X,Z}$ [82]. If X and Y are not homeomorphic, the natural pseudo-distance is set equal to ∞. In this way, two objects are considered as having the same shape if and only if they share the same shape properties, i.e., the natural pseudo-distance between the associated size pairs vanishes.

The *natural pseudo-distance* d_{np} is defined as the minimization of the change in measuring functions due to the application of homeomorphisms between topological spaces and it is related to the comparison of two shapes through persistence theory (see chapter 11).

CHAPTER 7

Algebraic Topology and Topology Invariants

Algebraic topology studies both topology spaces and functions through algebraic entities, such as groups or homomorphisms, by analyzing the representations (formally known as *functors*) that transform a topological problem into an algebraic one, with the aim of simplifying it. Then, the focus of algebraic topology is on the translation of the original problem into the algebraic language.

A fundamental contribution of algebraic topology is its support to the formal definition of a digital model and its description. On the one hand, the concepts of simplicial and cell complexes support the definition of a computational representation scheme consistent with the mathematical idealization of a shape; indeed cell decompositions are the most common geometric model used in computer graphics and CAD/CAM [134, 147]. On the other hand, homology groups analyze and classify smooth manifolds and complexes; for this reason we think that homology offers a set of mathematical tools (such as Betti numbers) that, possibly coupled with other descriptions, yield a synthetic and expressive shape description. Homology is a powerful tool for shape analysis, also because efficient algorithms for its computation exist. We refer the reader to well-known textbooks such as [99, 135, 149, 173] for a detailed treatment of other topics of algebraic topology.

7.1 CELL DECOMPOSITIONS

In order to construct topological spaces, one can take a collection of simple elements and glue them together in a structured way. Probably the most relevant example of this construction is given by simplicial complexes, whose building blocks are called simplices.

A *k-simplex* Δ^k in $\mathbb{R}^n$, $0 \leq k \leq n$, is the convex hull of $k+1$ affinely independent points $A_0, A_1, \ldots, A_k$, called *vertices*. Figure 7.1 shows the simplest examples of simplices: Δ^0 is a point, Δ^1 an interval, Δ^2 a triangle (including its interior), Δ^3 a tetrahedron (including its interior).

A k-simplex can be oriented by assigning an ordering to its vertices: two orderings of the vertices that differ by an even permutation determine one and the same orientation of the k-simplex. In this way, each k-simplex with $k > 0$ can be given a positive or a negative orientation. The *oriented k-simplex* with ordered vertices $(A_0, A_1, \ldots, A_k)$ is denoted by $[A_0, A_1, \ldots, A_k]$, whereas the k-simplex with opposite orientation is denoted by $-[A_0, A_1, \ldots, A_k]$.

A *face* of a k-simplex Δ^k is a simplex whose set of vertices is a non-empty subset of the set of vertices of Δ^k. A *finite simplicial complex* can now be defined as a finite collection of simplices

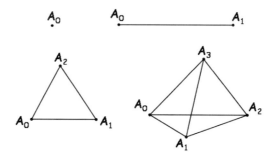

Figure 7.1: Examples of simplices Δ^0, Δ^1, Δ^2 and Δ^3 [24].

that meet only along a common face and their faces of any dimension. A concrete example of a simplicial complex is given by triangulated surfaces, where the vertices, edges and faces of the triangulation are 0-, 1- and 2-simplices, respectively. The *dimension* of a simplicial complex is the maximum dimension of its simplices.

A *subcomplex* of a complex K is a simplicial complex whose set of simplices is a subset of the set of simplices of K. Particular instances of subcomplexes are given by the star and the link of a simplex. Given a simplex Δ, the *star* of Δ is the union of all the simplices containing Δ. The *link* of Δ consists of all the faces of simplices in the star of Δ that do not intersect Δ. The concepts of star and link are illustrated in Figure 7.2 for the case of a 0-simplex. Other useful subcomplexes of a complex K are its skeletons: for $0 \leq r \leq \dim(K)$, the *r-skeleton* of K is the complex of all the simplices of K whose dimension is not greater than r.

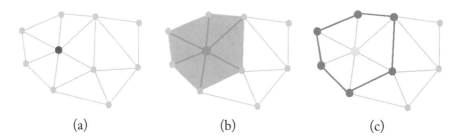

(a)   (b)   (c)

Figure 7.2: (a) A 0-simplex, (b) its star and (c) its link [24].

Note that it is also possible to define an *abstract simplicial complex* without using any geometry, as a collection $\mathcal{A}$ of finite non-empty sets such that if A is any element of $\mathcal{A}$, so is every non-empty subset of A. Simplicial complexes can be seen as the geometric realization of abstract simplicial complexes.

For more details on simplicial complexes refer to [149].

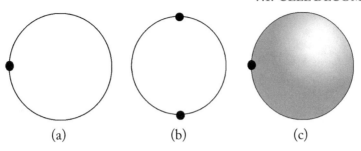

Figure 7.3: Examples of cell complexes. (a) Circle built starting with a point and attaching a 1-cell. (b) Circle built starting with two points and attaching two 1-cells. (c) Sphere built starting with a point and attaching a 2-cell [24].

Cell complexes   As simplicial complexes, *cell complexes* are constructed from basic building blocks, called cells, that generalize the concept of simplices. A λ-*cell* e^λ corresponds to the closed unit ball $B^\lambda = \{x \in \mathbb{R}^\lambda \mid \|x\| \leq 1\}$ of dimension λ. Cells are glued together via *attaching maps*. Attaching the cell e^λ to a space Y by the continuous map $\varphi : S^{\lambda-1} \to Y$, with $S^{\lambda-1} = \{x \in \mathbb{R}^\lambda \mid \|x\| = 1\}$ the boundary of B^λ, requires taking $Y \bigcup B^\lambda$, where each point $x \in S^{\lambda-1}$ is identified with the point $\varphi(x) \in Y$. The space so obtained is denoted by $Y \bigcup_\varphi e^\lambda$. It is important to note that different attaching maps φ can lead to different spaces.

The space X obtained by subsequently attaching finitely many cells is a finite *cell complex*. This means that there exists a finite nested sequence $\emptyset \subset X_0 \subset X_1 \subset \ldots \subset X_k = X$ such that, for each $i = 1, 2, \ldots, k$, X_i is the result of attaching a cell to X_{i-1}. Further details can be found in [90].

Examples of cell complexes are given in Figure 7.3 where the same circle is constructed through cell adjunction in two different ways. In (a), we start with a 0-cell, i.e., a point, and we connect a 1-cell via the attaching map that identifies the boundary of B^1 with the starting point. In (b), two 1-cells are attached to two 0-cells. In Figure 7.3(c) the sphere is obtained by attaching a 2-cell directly to a 0-cell, that is, identifying the boundary of B^2 with a point.

7.1.1   CONCEPTS IN ACTION

Data representation   From a historical perspective, the first type of model used in computer graphics and CAD/CAM was the *wireframe* model, which consists of the representation of edge curves and points on the object boundary. This incomplete model evolved further into *surface* or *boundary* models, that provide an unambiguous representation of the geometry of the 3D shape boundary, and into *solid* models, that encode a shape as a composition of volumes. The corresponding computational models generally imply the discretization of a shape into a simplicial or a cell complex. A model is called *simplicial* if its domain is discretized by a simplicial complex, while it is called *regular* if the domain is discretized through a *regular grid*, i.e., a cell complex

Figure 7.4: From left to right a cell decomposition [78], a triangle mesh, and a voxel grid.

in which all the cells are hypercubes. Triangle meshes are the most popular example of simplicial models. Examples of regular models are 2D and 3D images, where the cells are known as pixels in 2D and voxels in 3D, respectively [116]. Simplicial meshes are usually based on a piecewise linear interpolation of the shape geometry. Regular grids define a step-wise or analytical approximation of the shape geometry, according to the type of interpolation associated with the hypercubes.

Embedding these decompositions in a Euclidean space of lowest dimension [203] is extremely important for practical data representation and computation. For instance, the *Delaunay complex* D_X is a geometric simplicial complex that is homotopy equivalent to a given subspace of $X \in \mathbb{R}^n$ [65]. As a particular case, every orientable triangulated surface can be piecewise-linearly embedded in $\mathbb{R}^3$ [130].

Figure 7.4 shows three examples of data representations, from left to right a cell decomposition, a triangle mesh and a voxel grid.

7.2   HOMOLOGY

Homology theory offers the theoretical background for translating the study of topological properties, such as the number of holes, cavities, etc., of a shape, into algebraic structures. Despite the apparent complexity of the algebraic machinery behind homology, the advantage of using it is mainly applicative: indeed, homology groups can be effectively computed and, differently from homotopy, efficient algorithms for their computation exist [149].

The homology of a space is an algebraic object which reflects the topology of the space, in some sense counting the number of holes. The *homology* of a space X is denoted by $H_*(X)$, and is defined as a sequence of groups $\{H_q(X) : q = 0, 1, 2, \ldots\}$, where $H_q(X)$ is called the *q-th homology group* of X. In the literature there are various types of homologies [188]; the one we are addressing here is (integer) *simplicial homology*, which is strictly related to the concept of simplicial complex.

Let K be a simplicial complex in $\mathbb{R}^n$. For each $q \geq 0$, a *q-chain* of K is a formal linear combination $\sum_i a_i \Delta_i$, of oriented q-simplices Δ_i, with integer coefficients a_i. Two q-chains $a = \sum_i a_i \Delta_i$ and $b = \sum_i b_i \Delta_i$ are added componentwise, that is to say $a + b = \sum_i (a_i + b_i) \Delta_i$.

We denote by $C_q(K)$ the group of q-chains of K with respect to the addition; for q larger than n or for q smaller than 0, we set $C_q(K)$ equal to the trivial group. On the group $C_q(K)$, we can define the *boundary operator* $\partial_q : C_q(K) \to C_{q-1}(K)$. This is defined as the trivial homomorphism if $q \leq 0$, while for $q > 0$ it acts on each q-simplex via:

$$\partial_q[A_0, A_1, \ldots, A_q] = \sum_{i=0}^{q} (-1)^i [A_0, A_1, \ldots, A_{i-1}, \hat{A}_i, A_{i+1}, \ldots, A_q]$$

where $[A_0, A_1, \ldots, A_{i-1}, \hat{A}_i, A_{i+1}, \ldots, A_q]$ is the $(q-1)$-simplex obtained by eliminating the vertex A_i; the boundary map ∂_q extends by linearity to arbitrary q-chains. In Figure 7.5, the boundary operator is evaluated on some elementary simplices. The arrows represent the orientation of the simplices. Notice that changing the orientation of the simplices implies a different result for the boundary operator.

A chain $z \in C_q(K)$ is called a *cycle* if the boundary operator sends z to zero, i.e., $\partial_q z = 0$; a chain $b \in C_q(K)$ is called a *boundary* if it is the image, through the boundary operator, of a chain of dimension greater by one, i.e., there exists $c \in C_{q+1}(K)$ such that $b = \partial_{q+1} c$. The sets of cycles and boundaries form two subgroups of $C_q(K)$:

$$Z_q(K) = \{z \in C_q(K) \mid \partial_q z = 0\} = \ker \partial_q$$

$$B_q(K) = \{b \in C_q(K) \mid b = \partial_{q+1} c, \ \ for \ some \ c \in C_{q+1}(K)\} = \mathrm{Im} \partial_{q+1},$$

where ker and Im denote the kernel and the image of the map, respectively. It holds that $B_q(K) \subseteq Z_q(K)$, since $\partial_q \partial_{q+1} = 0$. The q-*th simplicial homology group* of K is then the quotient group:

$$H_q(K) = Z_q(K)/B_q(K).$$

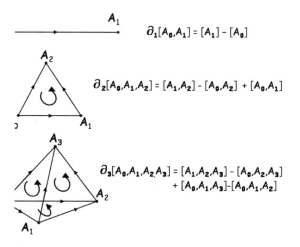

Figure 7.5: The boundary operator on elementary q-simplices [24].

Specifically, an element of $H_q(K)$ is an equivalence class, called *homology class*, of homologous q-cycles, i.e., cycles whose difference is a boundary. The homology $H_*(K)$ is a topological invariant of K; it is indeed an invariant of homotopy type.

The rank of $H_q(K)$ is called the *q-th Betti number* of K, and it is a measurement of the number of different holes in the space K. As an example, for three-dimensional data the three Betti numbers β_0, β_1 and β_2 count the number of connected components, tunnels and voids, respectively. The Betti numbers can be used to define a well-known topological invariant, the *Euler characteristic*: $\chi(K) = \sum_{i=0}^{n}(-1)^i \beta_i(K)$. Through the Euler characteristics it is possible to provide an alternative definition of the genus of a closed surface (cf. the definition in chapter 3) through the relationship $\chi = 2 - 2g$, where g is the genus. For surfaces with b boundary components, the relation becomes $\chi = 2 - 2g - b$.

A *simplicial map* f between two simplicial complexes K and L (i.e., a function from the set of vertices of K to the set of vertices of L such that the image of any simplex in K is a simplex in L) induces uniquely a set $H_*(f)$ of homomorphisms between the homology groups:

$$H_q(f) : H_q(K) \to H_q(L)$$

for each degree q. The maps $H_q(f)$ satisfy two elementary properties: (i) the identity map $id_K : K \to K$ induces the identity map on homology; and (ii) the composition $g \circ f$ of two maps corresponds to the composition $H_q(g \circ f) = H_q(g) \circ H_q(f)$ of the induced homomorphisms.

7.2.1   CONCEPTS IN ACTION

Betti numbers, homology groups and generators   In case of a triangle mesh T the Euler characteristics reduces to $\chi(T) = V - E + F$, where V, E and T represent respectively the number of vertices (0-simplices), edges (1-simplices) and faces (2-simplices) of the triangulation T. Therefore, its computation is straightforward. A recent example of application of Euler characteristics applied to the analysis and filtering of the seafloor has been proposed in [77].

More in general, there is a classical algorithm to compute the homology groups of an arbitrary simplicial complex (see Munkres [149]), based on reducing certain matrices to a canonical form, called *Smith normal form*. Unfortunately, the reduction of a matrix to its Smith normal form has a worst-case upper bound that is computationally prohibitive. To avoid the bottleneck of its computation, Delfinado & Edelsbrunner [56] describe an algorithm that computes the Betti numbers of a simplicial complex in $\mathbb{R}^3$. The intuition behind such an approach is to assemble the complex incrementally simplex by simplex, and at each step update the Betti numbers of the current complex. The algorithm runs optimally in time and space linear in the size of the complex.

For a triangulated surface of genus g, with a total of n simplices, a set of canonical generators of the homology groups can be computed in $O(gn)$ time, which is optimal in the worst case [204]. Each of the g or $2g$ canonical generators is represented by a polygonal curve whose vertices are on the 1-skeleton, while the other points are in the interior of a 2-simplex. A simple, yet computationally efficient, algorithm is to remove iteratively simplices preserving at each step the

Euler characteristics [93]. Unfortunately these generators are not optimal (in the sense of the shortest length) and in some cases the total number of edges of a single generator is $O(n)$ [58, 203].

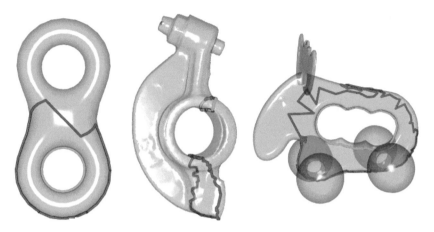

Figure 7.6: Homologous cycles of a set of models, the optimal cycles extracted by [59] are in green.

In general, computing the optimal homologous cycles under $\mathbb{Z}_2$ coefficients is NP-hard. The solution proposed in [59] is to switch to the coefficient group $\mathbb{Z}$. In that case, the problem of finding optimal cycles in a given homology class is reduced to an optimization problem, which is inherently an integer programming problem that can be solved in polynomial time. One consequence of this result is that it is possible to compute in polynomial time an optimal 2-cycle in a given homology class for any finite simplicial complex embedded in $\mathbb{R}^3$. Figure 7.6 depicts in red a set of homologous cycles and in green the optimal cycles obtained with the method [59].

The direct application of the homologous generators is to use them as the cuts to flatten a shape into a disk [75] and, in general, to parametrize it [93]. Recent advances on the approximation of loops in a computationally efficient way take advantage of topological persistence [60] and Reeb graphs [57], topics that are described in chapters 9 and 11.

CHAPTER 8

Differential Topology and Shape Analysis

Morse theory can be seen as the investigation of the relation between functions defined on a manifold and the shape of the manifold itself. The key feature in Morse theory is that information on the topology of the manifold is derived from the information about the critical points of real functions defined on the manifold. Let us first introduce the definition of Morse function, and then state the main results provided by Morse theory for the topological analysis of smooth manifolds, such as surfaces. A basic reference for Morse theory is [141], while details about notions of geometry and topology can be found, for example, in [104].

8.1 CRITICAL POINTS AND MORSE FUNCTIONS

Let M be a smooth compact n-dimensional manifold without boundary, and $f : M \to \mathbb{R}$ a smooth function defined on it. Then, a point p of M is a *critical point* of f if we have

$$\frac{\partial f}{\partial x_1}(p) = 0, \frac{\partial f}{\partial x_2}(p) = 0, \ldots, \frac{\partial f}{\partial x_n}(p) = 0,$$

with respect to a local coordinate system $(x_1, \ldots, x_n)$ about p. A real number is a *critical value* of f if it is the image of a critical point. Points (values) which are not critical are said to be *regular*. A critical point p is *non-degenerate* if the determinant of the *Hessian* matrix of f at p

$$H_f(p) = \left(\frac{\partial^2 f}{\partial x_i \partial x_j}(p) \right)$$

is not zero; otherwise the critical point is *degenerate*. Figure 8.1 shows some examples of non-degenerate and degenerate critical points. For a non-degenerate critical point p, the number of negative eigenvalues of the Hessian $H_f(p)$ of f at p determines the *index* of p, denoted by $\lambda(p)$.

We say that $f : M \to \mathbb{R}$ is a *Morse function* if all its critical points are non-degenerate.

A Morse function f is extremely simple near each non-degenerate critical point p. Indeed, we can choose appropriate local coordinates $(x_1, \ldots, x_n)$ around p, in such a way that f has a quadratic form representation: $f(x_1, \ldots, x_n) = f(p) - \sum_{i=1}^{\lambda(p)} x_i^2 + \sum_{i=\lambda(p)+1}^{n} x_i^2$.

Intuitively, the index of a critical point is the number of independent directions around the point in which the function decreases. For example, on a 2-manifold, the indices of minima, saddles, and maxima are 0, 1, and 2, respectively.

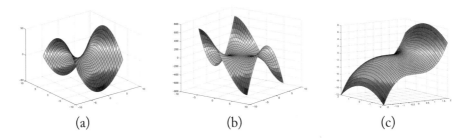

(a)   (b)   (c)

Figure 8.1: (a) The graph of $f(x, y) = x^2 - y^2$. The point $(0, 0)$ is a non-degenerate critical point. (b) and (c) The graphs of $f(x, y) = x^3 - 3xy^2$ (a "monkey saddle") and $f(x, y) = x^3 - y^2$. In both cases the point $(0, 0)$ is a degenerate critical point [24].

An important property is that a Morse function defined on a compact manifold admits only finitely many critical points, each of which is isolated. This means that, for each critical point p, it is always possible to find a neighborhood of p not containing other critical points.

8.1.1   INTEGRAL LINES

Given a Riemannian metric on M (see [79, 80, 94] for more details in these concepts) and a local coordinate system $(x_1, \ldots, x_n)$ with orthonormal tangent vectors $\frac{\partial}{\partial x_i}(P)$, $i = 1, \ldots, n$, the *gradient* of a function f in a point P is the vector:

$$\nabla f(P) = [\frac{\partial f}{\partial x_i}(P)]^T.$$

In particular the gradient is the zero vector iff the point P is critical.

Definition 8.1   *(Integral line)* An *integral line* $\gamma : \mathbb{R} \to M$ is a maximal path such that:

$$\frac{\partial \gamma}{\partial s}(s) = \nabla f(\gamma(s)), \ \forall s \in \mathbb{R}.$$

This means that the velocity vectors along the curve γ agree with its gradient. Each integral line is open at both ends and those points are critical points. Integral lines are pairwise disjoint and supposing that a critical point is an integral line itself, the integral lines partition M.

8.1.2   CONCEPTS IN ACTION

Critical points for shape characterization   Critical points and their configuration give a suitable framework to formalize and solve several problems related to shape understanding. For example, the surface characterization driven by the critical points of the height function has found several applications in the analysis of terrain modeling [9]. Figure 8.3 represents the critical points over a mathematical surface and a triangle mesh representing a terrain-like surface; red points correspond to maxima, blue points to minima and green points are saddles.

The computation of critical points on discretized surfaces received considerable attention in the literature. Banchoff [11] introduced critical points for height functions defined over polyhedral surfaces, by using a geometric characterization of critical points. A simplicial model in which linear interpolation is used on the triangles of the underlying mesh is the most common example of a polyhedral surface. Starting from the observation that a small neighborhood around a local maximum or minimum never intersects the tangent plane, as shown in Figure 8.2(a), while a similar small neighborhood is split into four pieces at non-degenerate saddles, as shown in Figure 8.2(b), the number of intersections is used to associate an *index* with each discrete critical point.

Consider the two-dimensional simplicial complex Σ in $\mathbb{R}^3$ with a manifold domain, and the height function $\xi : \mathbb{R}^3 \to \mathbb{R}$ with respect to the direction ξ in $\mathbb{R}^3$; ξ is called *general for* Σ if $\xi(v) \neq \xi(w)$ whenever v and w are distinct vertices of Σ. Under these assumptions, critical points may occur only at the vertices of the simplices and the number of times that the plane through vertex p and perpendicular to ξ cuts the link of p is equal to the number of 1-simplices in the link of p with one vertex above the plane and one below (see Figure 8.2). Point p is called *middle* for ξ for these 1-simplices. Then, an indexing scheme is defined for each vertex of Σ as follows [10]:

$$i(v, \xi) = 1 - \frac{1}{2}(number\ of\ 1 - simplices\ with\ v\ middle\ for\ \xi). \qquad (8.1)$$

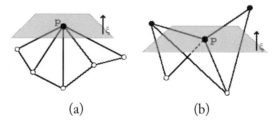

(a)   (b)

Figure 8.2: Configuration of vertices around a maximum point (a) and a non-degenerate saddle (b) [24].

Discrete critical points are at the vertices of the simplicial model and are defined as points with index different from 0. In particular, the index is equal to 1 for maxima and minima, while it can assume an arbitrary negative integer value for saddles.

The characterization provided by Banchoff correctly distinguishes critical points in dimension 2 and 3. For higher-dimensional spaces the Betti numbers of the *lower link*, that is the set of connected components of the link of a vertex which join points with a height less than that the vertex, provide a more complete characterization of discrete critical points, as suggested in [66].

It has to be observed that, since the definition of critical points has a local nature, small perturbations of the shapes can considerably influence the number of critical points detected. Moreover, in many applications the shapes to be analyzed are likely to have degenerate critical points. A simple solution is to locally simulate the "simplicity" of the critical points [73] locally perturbing the surface but in general it is necessary to further filter and structure them in a hierarchical structure, as explained in the descriptions and theories described in the chapters 9, 10 and 11.

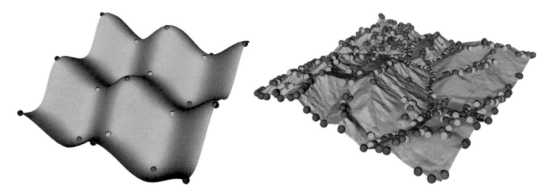

Figure 8.3: Critical points for a mathematical surface and a terrain-like surface. Red points correspond to maxima, blue points to minima, and green points are saddles.

8.2 TOPOLOGICAL ANALYSIS THROUGH (LOWER) LEVEL SETS

Topological information about M is captured by the changes of the level sets and the lower level sets of M relative to the function f. The *level set* of f corresponding to the real value t is the set of points $V_t = \{p \in M \mid f(p) = t\} = f^{-1}(t)$; t is called an *isovalue* of f. The *lower level set* is given by $M_t = \{p \in M \mid f(p) \leq t\} = f^{-1}((-\infty, t])$.

We begin by studying how the lower level set M_t changes as the parameter t changes (Figure 8.4(a)). Morse theory states that the topology of M_t stays unchanged as the parameter t goes through regular values of f, while changes occur when t passes through a critical value. More precisely, the following theorem holds:

Theorem 8.2   *Let a, b be real numbers such that $a < b$ and the set $f^{-1}([a, b])$ contains no critical points for f. Then M_a and M_b have the same topology.*

Actually, a stronger result holds in this case stating that M_a and M_b are diffeomorphic, namely, there is a differentiable and invertible function between M_a and M_b, whose inverse is also differentiable (see Figure 8.4(b)).

Theorem 8.3   *Let p be a critical point of f with index λ and let $f(p) = c$. Then, for each ε such that $f^{-1}([c - \varepsilon, c + \varepsilon])$ contains no critical points other than p, the set $M_{c+\varepsilon}$ has the same homotopy type of the set $M_{c-\varepsilon}$ with a λ-cell attached:*

$$M_{c+\varepsilon} \cong M_{c-\varepsilon} \cup_{\varphi_p} e^\lambda.$$

According to the definitions in section 7.1, the *attaching map* φ_p identifies each point $x \in S^{\lambda-1}$ with the point $\varphi_p(x) \in M_{c-\varepsilon}$ (see Figure 8.4(c)).

In order to study the changes in the level sets $V_t = f^{-1}(t)$, an approach to Morse theory based on the attaching of handles [142], rather than cells, can be used, as in [89] for the case of surfaces. When f is defined on a surface, if t is a regular value for f then V_t, if not empty, is the union of finitely many smooth circles. Moreover, if a, b are real numbers such that $a < b$, then

1. if the set $f^{-1}([a, b])$ contains no critical points for f, then V_a and V_b are diffeomorphic;

2. if the set $f^{-1}([a, b])$ contains only one critical point of index 0 for f, then V_b is the union of V_a with a circle;

3. if the set $f^{-1}([a, b])$ contains only one critical point of index 2 for f, then V_b is diffeomorphic to V_a without one of its circles;

4. if the set $f^{-1}([a, b])$ contains only one critical point of index 1 for f, then the number of connected components of V_b differs from that of V_a by -1, 0 or 1 depending on the attaching map.

In case (4), the difference in the number of connected components is non-zero if the handle (in this case, the strip $[0, 1] \times [0, 1]$) is attached without twists (or with an even number of twists), while it is 0 if there is an odd number of twists. The presence of an odd number of twists implies that the surface is non-orientable. Therefore, when the surface is embedded in $\mathbb{R}^3$, V_a and V_b necessarily have a different number of connected components.

8.3   HOMOLOGY OF MANIFOLDS

Morse theory asserts that changes in the topology of a manifold endowed with a Morse function occur in the presence of critical points; since most manifolds can be triangulated as simplicial complexes and a Morse function can be discretized on simplices, those changes in the topology

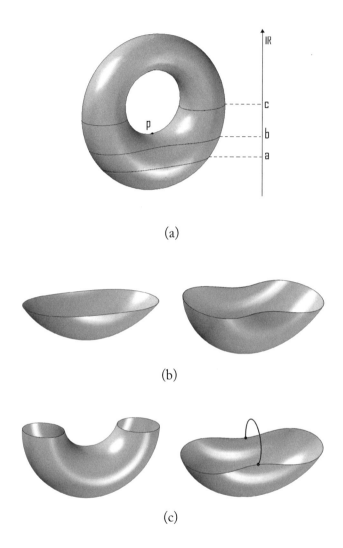

Figure 8.4:   (a) A manifold M and three lower level sets M_a, M_b, M_c, with respect to the height function. (b) There are no critical points in $f^{-1}([a, b])$: M_a (left) and M_b (right) are diffeomorphic. (c) The passage through the critical point p of index 1 causes a topological change: M_c (left) has the same homotopy type as M_b with a 1-cell attached (right) [24].

can be interpreted in terms of homology. Thus, we have the following description of the homology of a manifold M [90]:

Theorem 8.4   *Let a, b be real numbers such that $a < b$ and $f^{-1}([a, b])$ contains only one critical point p of f, of index λ, and let φ_p be the attaching map of the λ-cell corresponding to p. Then:*

(a) if $k \neq \lambda$ and $k \neq \lambda - 1$ then $H_k(M_b) \cong H_k(M_a)$

(b) $H_{\lambda-1}(M_b) \cong H_{\lambda-1}(M_a)/Im H_{\lambda-1}(\varphi_p)$

(c) $H_\lambda(M_b) \cong H_\lambda(M_a) \oplus Ker H_{\lambda-1}(\varphi_p)$

This means that, depending on the attaching map φ_p, only the homology degrees $\lambda - 1$ and λ can be affected by the adjunction of a λ-cell. In particular, the Betti number $\beta_{\lambda-1}$ can decrease and β_λ can increase. For example, in the case shown in Figure 8.4, when we pass through the critical point p according to the increasing value of f, a tunnel is created, and β_1 increases from the value 0 to the value 1.

This characterization gives a hint of the ideas underlying Reeb graphs, size theory, topological persistence and Morse shape descriptors, detailed in chapters 9, 10 and 11.

CHAPTER 9

Reeb Graphs

Reeb graphs encode the evolution and the arrangement of the level sets of a real function defined on a shape. They were first defined by a French mathematician, Georges Reeb, in 1946 [166] as topological constructs. In recent years, Reeb graphs have become popular in computer graphics as tools for shape description, analysis and comparison.

Reeb graphs present a framework for studying a shape on the basis of an arbitrarily chosen real function. They effectively code both topological and geometrical information; the topological analysis is driven by the evolution of the level sets of the function chosen and the geometry properties reflect the shape of the contours. Different functions give insights on the shape from different perspectives and the properties of the resulting graphs are parametric to these functions.

9.1 REEB GRAPH DEFINITION

Given a manifold M and a real-valued function f defined on M, the simplicial complex defined by Reeb, conventionally called the Reeb graph, is the quotient space defined by the equivalence relation that identifies the points belonging to the same connected component of level sets of f. Under some hypotheses on M and f, Reeb stated the following theorem, which actually defines the Reeb graph.

Theorem 9.1 *Let M be a compact n-dimensional manifold and f a simple[1] Morse function defined on M, and let us define the equivalence relation "$\sim$" as $(P, f(P)) \sim (Q, f(Q))$ iff $f(P) = f(Q)$ and P, Q are in the same connected component of $f^{-1}(f(P))$. The quotient space on $M \times \mathbb{R}$ induced by "$\sim$" is a finite and connected simplicial complex K of dimension 1, such that the counter-image of each vertex Δ_i^0 of K is a singular connected component of the level sets of f, and the counter-image of the interior of each simplex Δ_j^1 is homeomorphic to the topological product of one connected component of the level sets by $\mathbb{R}$ [74, 166].*

Reeb also demonstrated the following theorems, which clarify the relations between the degree, or order, of the vertices of the simplicial complex K associated with the quotient space and the index of the corresponding critical points.

Theorem 9.2 *The degree of a vertex Δ_i^0 of index 0 (or n) is 1 and the index of a vertex Δ_i^0 of degree 1 is 0 or n.*

[1]A function is called simple if its critical points have different values.

Theorem 9.3   *If $n \geq 3$ the degree of vertices Δ_i^0 of index 1 (or $n - 1$) is 2 or 3. If $n = 2$ the degree of vertices Δ_i^0 of index 1 is 2, 3, or 4. The degree of vertices Δ_i^0 of index different from 0, 1, $n - 1$, or n is 2.*

In other words, the theorem 9.2 states that leaf nodes of K can be either maxima or minima of f, while from the theorem 9.3 we can deduce that, for 2-manifolds that can be embedded in $\mathbb{R}^3$, the degree of vertices representing saddles is always 3 (see section 8.1).

Several variations/approximations of the Reeb graph exist; among the others we mention the Multiresolution Reeb Graph (MRG) [102], the Extended Reeb Graph (ERG) [6], the augmented Multiresolution Reeb Graph (aMRG) [202], the Discrete Reeb Graph (DRG) [211].

Strictly related to Reeb graphs are *contour trees*. Contour trees are a special case of Reeb graphs, in which the shape (i.e., the domain of the function) is simply connected and with a single border element. Contour trees were introduced for practical issues, mainly as a tool for topographic analysis. Indeed, the evolution of contour lines on a surface explicitly represents hills and dales with their elevation-based adjacency relationships. Therefore, contour trees are an efficient data structure to store containment relationships among contours in contour maps, typically representing terrain elevations [30]. Several variations of contour trees have been proposed in the literature, as for example the *volume skeleton tree* by Takahashi et al. [192]. The contour tree may also be augmented with further information on all topological changes of the level sets by adding nodes that correspond to critical values where not the number but the topology of the contours changes. Examples are the *contour topology tree* [46], the *criticality tree* [52], and the *topographic change tree* [87].

Computational complexity   Historically, the first algorithm was proposed by Shinagawa and Kunii in 1991. It ran over height functions defined on a triangulated 2-manifold [181] and had a computational complexity of $O(n^2)$, where n is the number of triangles in the triangulation. The algorithm proposed by Cole-Mclaughlin et al. [48] showed how to compute the graph in $O(n \log n)$ operations by maintaining the level sets using dynamically balanced search trees following the algorithm for contour trees in all dimensions by Carr et al [40]. However, in higher dimensions, the presence of loops in the Reeb graph implies that its decomposition into a join and split tree, which was crucial for the efficiency of the algorithm by Carr et al. [40] and Cole-Mclaughlin et al. [48], may not exist and, therefore the algorithm proposed in [48] does not extend to 3-manifolds.

Pascucci et al. [159] proposed an online algorithm that constructs the Reeb graph for streaming data. Their algorithm takes advantage of the coherency in the input to construct the Reeb graph for simplicial complexes of arbitrary dimension. Though the algorithm performs well in practice, it has a worst case time complexity of $O(nm)$, where n is the number of vertices and m is the number of triangles in the complex. Tierny et al. further improved the complexity of computing a Reeb graph of $3D$ manifolds with boundary embedded in R^3 using cuts to reduce the extraction to the computation of a set of contour trees reducing the time complexity to $O m \log m + h m)$, where h is the number of independent loops in the Reeb graph.

Doraiswamy and Natarajan utilized an efficient tree-cotree [63] decomposition-based representation of level sets to construct the Reeb graph in $O(m g \log m)$ time, where m is the number of triangles in the tetrahedral mesh representation of a 3-manifold, and g is the maximum genus over all level sets of the function. Using an efficient representation of the tree-cotree partition results in an improved $O(m \log m + m \log g (\log \log g)^3)$ time algorithm that for d-manifolds reduces to $O(m \log m (\log \log m)^3)$ operations.

Harvey et al. [98] presented the first sub-quadratic algorithm to compute the Reeb graph for an arbitrary simplicial complex. That algorithm proposes a randomized access to the data with an expected running time $O(m \log n)$, where m is the size of the 2-skeleton of the complex (i.e., total number of vertices, edges and triangles), and n is the number of vertices. Finally, Parsa [157] gave the first deterministic algorithm able to compute the Reeb graph in $O(m \log m)$ operations (m represents the size of the 2-skeleton), which is optimal if the number of edges of the complex is in the same order as the number of vertices.

9.2   REEB GRAPHS ON 2- AND 3-MANIFOLDS

As a consequence of its ability to extract high-level features from shapes, the Reeb graph is an effective tool for shape analysis and description, especially in case of 2- and 3-manifolds. Figure 9.1 shows two examples of Reeb graphs of a closed surface. In Figure 9.1(a) some level sets of the height function are drawn with the corresponding Reeb graph; in Figure 9.1(b) the Reeb graph of the same object is shown using the Euclidean distance from a point.

(a)   (b)

Figure 9.1: Reeb graphs with respect to the height function (a) and the distance from the point p (b) [24].

For orientable, closed 2-manifolds, the number of cycles in the Reeb graph corresponds to the genus of the manifold, and this result has been generalized in [48], where the authors demonstrate that the number $\beta_1(M)$ of non-homologous loops of the surface is an upper bound of the number of loops $\beta_1(K)$ of the Reeb graph. The equality holds in case of orientable surfaces without boundary [48], while, in general, the following relation is satisfied:

$$g \leq \beta_1(K) \leq 2g + b_M - 1,$$

where b_M denotes the number of boundary components of the 2-manifold M having genus g. Theoretical results are available for non-orientable 2-manifolds. In this case, the number of loops of the Reeb graph verifies the following relations: $0 \leq \beta_1(K) \leq \frac{g}{2}$ when M is closed, and $0 \leq \beta_1(K) \leq g + b_M - 1$ for manifolds with boundary. As for 3-manifolds, it is not true that the number of the loops of an orientable, closed 3-manifold is independent of the function f. In addition, it has been proven that for every 3-manifold M there exists at least one Morse function f such that the Reeb graph of M with respect to f is a tree [48].

For 3-manifolds with boundary, the 3-manifold structure is studied either by introducing a virtual closure of the manifold [68], or by associating a Reeb graph to each 2-manifold boundary component of the 3-manifold and keeping track of the changes between interior and void with a supplementary graph [180, 181]. Several techniques have been reported in the literature for the computation of Reeb graphs on $2D$ and $3D$ models [25], but also for higher dimensions [63, 98, 157, 159, 196].

9.3   CONCEPTS IN ACTION

Feature selection and Reeb graphs for 3D classification   Classifying 3D shapes is an important issue for many applications. In [28] the authors propose a scheme for supervised 3D classification in a multi-class problem, having two main ingredients: the first is the description of 3D objects via spectral features obtained from Reeb graphs; the second is a feature selection technique based on information theory. More precisely, three types of Reeb graphs are extracted from each 3D object, corresponding to three real functions (Fig. 9.2, left). For each graph, nine different measures are calculated and transformed into histograms (Fig. 9.2, middle). Only structural graph information is taken into account, that is, only the graph connectivity, without any geometric attribute. Then, the feature selection process decides which measure and which graph to discard (Fig. 9.2, right); this is done by exploiting information theory to filter those features which do not contribute to the separability of the objects into classes.

The results in [28] demonstrate the feasibility of multi-class 3D classification based on unattributed graphs and purely structural spectral features. Moreover, the authors investigate the importance of each of the spectral features used, and to what extent that importance depends on the real function used to extract Reeb graphs.

Combining topological graphs and geometric attributes for efficient shape retrieval   Among the different methods for retrieval using Reeb graphs, we describe here the method in [12] that associates geometric attributes to the Extended Reeb graph [6] and performs 3D shape retrieval through a graph matching technique based on the combination of several kernels. In such an approach the model parts are coupled to nodes and edges: each node n of a graph G represents one region $r \in R(S)$ and each edge $e = (v_1, v_2)$ is associated with the surface portion bounded by the regions r_1 and r_2 associated with v_1 and v_2, respectively. The geometric descriptor selected to code the model parts is the spherical harmonic descriptor [84]. Even if any feature vector de-

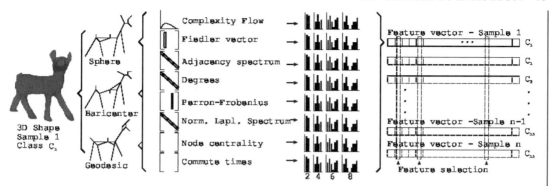

Figure 9.2: From left to right, 3D description via Reeb graph extraction, computation of histograms of spectral features, feature selection [85].

scriptor could be used to code the model subparts, this descriptor satisfies the requirements of being rotation and scale invariant, computationally efficient, synthetic and able to store the distribution of each sub-part. Nodes and edges are uniformly indexed using an array, whose entries correspond on the spherical harmonic values of the related sub-part.

Figure 9.3 shows a set of contours and the Reeb graph of a tea-pot model with the corresponding shape parts. In case of multiple connection between the nodes v_1 and v_2, each edge stores the geometric information related to a single portion of the surface.

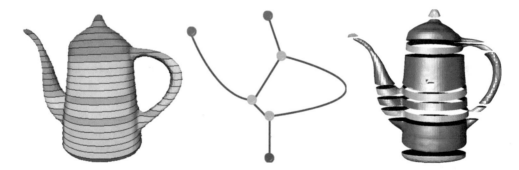

Figure 9.3: Characterization, ERG and model parts associated with graph nodes and edges.

Several methods are available to measure the similarity between graphs G and G', most of them derived from graph matching techniques [201]. The approach followed in [12] adopts a matching techniques based on kernels [109, 119, 206, 207].

The definition of the ERG is flexible with respect to the choice of the function f: this property makes the ERG adaptable to the characteristics of the specific domain where the similarity needs to be evaluated. Therefore, given a 3D model database composed of N models, a $N \times F$ graph table T is created (being F the number of real functions). The element $T(i, j)$ is the ERG

j associated to model i. Finally, the combination of the different structural/feature-based kernels is used to enhance the retrieval results. Since each real function f_j provides a similarity vector for the query q, the first r nearest neighbors of q seen with f_j can be computed by ranking the similarities in decreasing order. The application of the F real functions thus gives F nearest neighbor lists for a query q, viewed with each of these functions. An aggregation procedure is adopted to define a single kernel, function of the basis kernels. In general, weights can encode a priori knowledge about, e.g., the kernel ability to retrieve some particular classes or expert knowledge. They can also be learned from a set of training examples [13, 51].

Reeb graphs for partial matching   Being a topological structure that explicitly associates the graph elements (nodes and edges) to shape parts, the Reeb graph has also been successfully adopted for partial correspondence [26, 197].

In [197] each shape is decomposed into charts with disk or annulus topology only and the partial similarity between two shapes is evaluated by computing a variant of their maximum common sub-graph. This shape decomposition enables the computation of concise and efficient signatures for object parts based on parameterization techniques (see Figure 9.4). To limit the

Figure 9.4: Reeb chart (bright colors) and pattern (dark colors) matching between a boy and a centaur. Unmatched charts are black.

number of possible combinations during the matching process, each signature of the decomposition (Reeb pattern) is matched only with topologically equivalent patterns. Since the Reeb graph extraction is based on the geodesic distance from shape extrema, the description is also invariant against rigid transformations and robust against non-rigid transformations and surface noise.

CHAPTER 10

Morse and Morse-Smale Complexes

The intuition behind Morse and Morse-Smale complexes was given by Maxwell [136]:

> "Hence each point of the earth's surface has a line of slope, which begins at a certain summit and ends in a certain bottom. Districts whose lines of slope run to the same bottom are called basins or dales. Those whose lines of slope come from the same summit may be called, for want a better name, hills. Hence, the whole earth may be naturally divided into basins or dales, and also, by an independent division, into hills, each point of the surface belonging to a certain dale and also to a certain hill."

The decomposition of the surface into its hills corresponds to the ascending, Morse complex. The decomposition of the surface into its dales corresponds to the descending, Morse complex. If we overlap the decompositions on hills and on dales, we obtain a Morse-Smale decomposition.

Morse and Morse-Smale complexes describe a shape by decomposing it into cells of uniform behavior of the gradient flow of a real function f applied on the shape, and by encoding the adjacencies among these cells in a complex which describes both the topology and the geometry of the gradient of f.

Morse and Morse-Smale complexes were introduced in computer graphics for the analysis of two-dimensional scalar fields and their use has been extended to handle three-dimensional scalar fields and more generic shapes. The theory behind is more general, however, with connections to the theory of dynamical systems [156]. The difference between Morse and Morse-Smale complexes concerns the properties of the function f which is used to study the shape, as it will be explained later on. Alternatively, this decomposition can be interpreted as having been obtained by joining the critical points of the function f by lines, in the case of a two-dimensional scalar field (or surfaces in the case of a three-dimensional scalar field), of steepest ascent or descent of the gradient.

10.1 BASIC CONCEPTS

Let M be a smooth compact n-manifold without boundary, and let $f : M \to \mathbb{R}$ be a smooth Morse function. Let us also assume that M is embedded in $\mathbb{R}^n$ or that a Riemannian metric is defined on M. Morse complexes are induced by the partition induced by the integral lines of the function f over M (see chapter 8, definition 8.1). Indeed, integral lines are pair-wise disjoint, that is, if their images share a point, then they are the same line. The images of integral lines cover the whole M, but if we consider the integral lines associated with the critical points of f, their images define a partition of M.

This partition is used to decompose M into regions of uniform flow, thus capturing the characteristics of the gradient field. More precisely, the *descending manifold* of a critical point p is the set $D(p)$ of points that flow towards p, and the *ascending manifold* of p is the set $A(p)$ of points that originate from p^1. In formulae:

$$A(p) = \{q \in M : \lim_{t \to +\infty} \gamma_q(t) = p\}$$

$$D(p) = \{q \in M : \lim_{t \to -\infty} \gamma_q(t) = p\},$$

where γ_q is the integral line at the point q. Note that the descending manifold of f is the ascending manifold of $-f$.

The descending manifold of a critical point p of index i is an open i-cell. Similarly, the ascending manifold of a critical point of index i is an $n - i$ open cell. For example, if M is a 2-manifold the descending manifold of a maximum is an open disk, that of a saddle is an open interval, and that of a minimum is the minimum itself.

The collection of all descending manifolds forms a complex, called the *descending Morse complex*, and the collection of all ascending manifolds also form a complex, called the *ascending Morse complex*, which is dual with respect to the descending complex (see Figure 10.1 left and middle). For instance, when M is a 2-manifold, the 2-cells of the descending Morse 2-complex correspond to the maxima of f, the 1 cells to the saddle points, and the 0-cells to the minima. Symmetrically, the 2-cells of the ascending Morse 2-complex correspond to the minima of f, the 1-cells again to the saddle points, and the 0-cells to the maxima. When M is a 3-manifold, the 3-cells of a descending Morse 3-complex correspond to the maxima, the 2-cells to the 2-saddles the 1-cells to the 1-saddles, and the 0-cells to the minima. Symmetrically, the 3-cells of the ascending Morse 3-complex correspond to the minima, the 2-cells to the 1-saddles, the 1-cells to the 2-saddles, and the 0-cells to the maxima.

The function f is a *Morse-Smale* function if the descending and ascending Morse complexes intersect only transversally.^{2} In 2D, this means that, if an ascending 1-manifold intersects a descending 1-manifold transversally, they cross at exactly one point. This condition implies that the topological behavior of the images of the integral lines does not change under small perturbations of the vector field [156].

In the case of Morse-Smale functions, the *Morse-Smale complex* is defined as the intersection of the ascending and descending manifolds. The cells of the *Morse-Smale complex* are the

---

^{1}In the mathematical literature, the term *unstable* is used instead of ascending, and the term *stable* is used instead of descending [156].

^{2}By definition, two submanifolds A and B of a manifold M intersect transversally in p if $T_p A + T_p B = T_p M$ where T_p is the tangent space at p.

components of sets $D(p) \cap A(q)$, for all critical points p and q of function f [69, 70]. Each cell of the Morse-Smale complex is the union of the integral lines sharing the same origin p of index i and the same destination q of index j. The dimension of the cell is given by the difference of the indices. Examples of descending and ascending manifolds on a surface and the resulting Morse-Smale complex are shown in Figure 10.1.

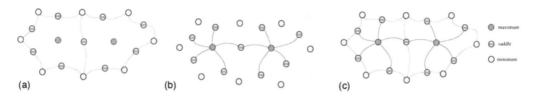

Figure 10.1: (a) Descending manifolds of maxima and saddles; (b) Ascending manifolds of minima and saddles; (c) The Morse-Smale complex given by the overlay of these ascending and descending manifolds.

The Morse-Smale complex is characterized by cells with a regular connectivity. In the 2D case, each saddle point p has four incident 1-cells, two joining p to maxima, and two joining p to minima. Such 1-cells alternate in a cyclic order around p. Also, the 2-cells are quadrangles whose vertices are critical points of f of index $1, 0, 1, 2$ (i.e., saddle, minimum, saddle, maximum) in this order. In the 3D case, all 2-cells are quadrangles whose vertices are a minimum, 1-saddle, 2-saddle, 1-saddle in this order (*quadrangles of type 1*), or a 1-saddle, a 2-saddle, a maximum, a 1-saddle in this order (*quadrangles of type 2*). A 1-cell connecting a 1-saddle and a 2-saddle is on the boundary of four quadrangles that alternate between quadrangles of type 1 and type 2. The 3-cells are called *crystals* and are bounded by quadrangles [69, 70].

The 1-skeleton of a Morse-Smale complex is a 1-complex formed by integral lines joining critical points. Similar structures among critical points have been widely studied in the literature under the name of *critical nets*. A graph representation of the critical net in a two-dimensional Morse-Smale complex is the so-called *surface network* [163, 174], widely used in spatial data processing for morphological terrain modeling and analysis (see [165] for an interesting collection of contributions on this specific topic).

Morse and Morse-Smale complexes have been extensively studied, mainly for the understanding and visualization of scalar fields, but also for more general applications in shape analysis, by using as function the curvature [133, 155], or the Connolly function [41]. A considerable number of algorithms has been developed for extracting critical points and lines, with a specific focus on terrain modeling and analysis. In general, region-based methods aim at extracting a Morse complex, while boundary-based approaches typically focus on the computation of a Morse-Smale complex. In case of simplicial models the most popular algorithms for the extraction of the Morse and Morse-Smale complexes are [31, 41, 54, 55, 69, 70, 97, 131, 151, 158].

10.2   CONCEPTS IN ACTION

Generalization of Morse-Smale complexes   Topology-based methods used for visualization and analysis of scientific data are becoming increasingly popular. Their main advantage lies in the capability to provide a concise and effective synthesis of important structures in scientific data sets, where interactive exploration of data is a means to understand phenomena at hand. Consider for instance, simulations of turbulent mixing between fluids, or in molecular shape analysis: the interactive exploration of data in this field requires the handling of huge, complex and possibly noisy data. Therefore, flexible hierarchical, or multi-resolution, representations are welcome in order to allow for an adaptive data simplification and for detecting important structures [32–34].

In the case of Morse-Smale complexes, it has been shown how multi-resolution representations can be obtained by successively coupling and eliminating pairs of critical points. For a 2D scalar field, the so-called *cancellation* consists of collapsing a maximum-saddle pair into a maximum, or a minimum-saddle pair into a minimum. A cancellation simulates the smoothing of the scalar field by modifying the gradient flows around two critical points. Figure 10.2 shows an example of cancellation on a surface network. The various techniques proposed in the literature differ in the criterion used to filter the critical points, and in the order they are eliminated. For example, in [32, 70], a saddle s is chosen together with its adjacent maximum at lower elevation, or its adjacent minimum at higher elevation; the order in which the pairs of points are cancelled is determined by their persistence, according to the principle that persistence is equivalent to significance (cf. chapter 11).

The case of 3D Morse-Smale complexes has been investigated in [96] and more recently in [49, 95, 97]. These methods extend 2D techniques to functions defined over 3-manifolds. The extension is not trivial, since in 3D there are three possible types of legal cancellations: minimum and 1-saddle, 1-saddle and 2-saddle, and 2-saddle and maximum. While the simplification involving any extremum (minimum or maximum) are similar to their 2D counterparts, the saddle-saddle cancellation does not have an analogue in lower dimensions. To ensure that minima and maxima originally separate by two saddles do not connect, additional cells have to be introduced in the Morse-Smale complex, which are removed by subsequent saddle-extremum cancellations as proposed in [96].

As an application of the multi-resolution generalization of Morse-Smale complexes, Bremer et al. [34] propose the rendering of a 1201-by-1201 single-byte integer value terrain data set of the Grand Canyon, and the identification of oil extraction sites in an underground oil reservoir, as local minima in the simplified Morse-Smale complex of oil pressure data.

In [33] two other applications are shown. One is molecular biology, where the focus is the segmentation of a molecular surface into cavities and protrusions. Given the skin surface of a protein complex and the atomic density function over this surface, the ascending manifolds of minima of this function segment the surface into protrusions; simplifying the Morse-Smale complex captures protrusions at coarser and coarser level, as shown in Figure 10.3.

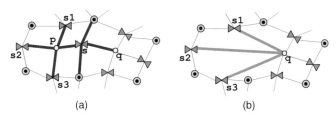

Figure 10.2: (a) A surface network $S_N = (C, A)$. The arcs involved in the cancellation are highlighted. (b) The surface network $S'_N = (C', A')$ obtained from S_N by eliminating the saddle s and the minimum p. The new arcs are highlighted.

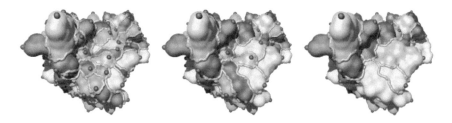

Figure 10.3: Segmentation of the atomic density function on a molecule in [33]. Minima are shown in blue and ascending paths in gold. Three segmentations into 198 (left); 100 (middle) and 50 (right) protrusions are depicted.

Another application is the physics simulation of the turbulent mixing between two fluids. In particular, scientists are interested in *bubbles* formed during the mixing process and their automatic segmentation. Given an iso-surface between two mixing fluids extracted from one time-step of a simulation and the z-coordinate as the Morse function, bubbles can be defined as the descending manifolds of maxima. Nevertheless, there exist many superfluous maxima caused by noise in the data set. As it happens with molecular surfaces, a uniform simplification of the MS complex helps remove most of these artifacts and create a much cleaner segmentation.

Finally, Morse and Morse-Smale complexes are strongly related to visualization of vector field topology [101, 194].

CHAPTER 11

Topological Persistence

Persistent homology couples notions of algebraic topology (see chapter 7) with Morse theoretic reasoning: as we have seen in chapter 8, Morse theory explores the topological attributes of an object in an evolutionary context. This concept of evolution has been rethought in [72], where the authors introduced *persistence*, a technique which grows a space incrementally and analyzes the placement of topological events within the history of this growth: for example, the *birth* of a connected component and its *death* when it merges into another component, the birth of a hole and its death after it is filled (section 11.1).

The aim was to furnish a scale to assess the relevance of topological attributes, under the assumption that *longevity* is equivalent to *significance*. In other words, a significant topological attribute must have a long lifetime in a growing complex, whereas noise and details are short-lived. In this way, one is able to distinguish the essential features from the fine details. The lifespan of topological attributes is encoded in a simple and compact representation called *persistence diagram*, which we describe in section 11.2. Their multidimensional analogues, namely *persistence spaces*, are the subject of section 11.3.

We notice that similarly to many concepts in mathematics, though the term *persistence* was introduced in [71] in 2000, the concept has a historical root system that comprises the work of Patrizio Frosini and collaborators, who in the 1990s introduced size functions [81], which are equivalent to 0-persistent homology, and the study of the homology of sample spaces by Vanessa Robins [170] that dates back to 1999.

Understanding this chapter requires the notions about simplicial complexes and homology introduced in chapter 7.

11.1 BASIC CONCEPTS

The first concept related to persistent homology theory is that of a filtered complex, that is a complex equipped with a filtration.

A *filtration* of a complex is a nested sequence of subcomplexes that ends with the complex itself. Formally, a complex K is filtered by a filtration $\{K^i\}_{i=0,\dots,n}$ if $K^n = K$ and K^i is a subcomplex of K^{i+1} for each $i = 0, \dots, n-1$.

An example of filtered complex is given in Figure 11.1 (top). Since the sequence of subcomplexes K^i is nested, one can think of K as a complex that grows from an initial state K_0 to a final state $K^n = K$. Therefore K is often referred to as a growing complex.

Filtered complexes arise naturally in many situations. The simplest example of filtration is the *age filtration*: the complex K is filtered by giving an ordering $\Delta_0, \Delta_1, \ldots, \Delta_m$ to its simplexes and by defining the sequence of its subcomplexes K^i as $K^i = \{\Delta_j \in K : 0 \leq j \leq i\}$. In other words the complex grows from $K^0 = \{\Delta_0\}$ adding each simplex one by one according to the given order.

A filtered complex also occurs when some space (e.g., a curve or a surface) is known only through a finite sample X of its points. Since the knowledge of the original space is necessarily imprecise, a multi-scale approach may be suited to describe the topology of the underlying space. The idea is to construct, for a real number $\epsilon > 0$, an abstract simplicial complex $R_\epsilon(X)$, called the *Rips complex*, whose abstract k-simplexes are exactly the subsets $\{x_0, x_1, \ldots, x_k\}$ of X such that $d(x_i, x_j) \leq \epsilon$ for all pairs x_i, x_j with $0 \leq i, j \leq k$. Whenever $\epsilon < \epsilon'$, there is an inclusion $R_\epsilon(X) \rightarrow R_{\epsilon'}(X)$ that reveals a growing complex.

Given a filtered complex, its topological attributes change through the filtration, since new components appear or connect to the old ones, tunnels are created and closed off, voids are enclosed and filled in, etc. In particular, as for 0-homology, each homology class corresponds to a connected component, and a homology class is born when a point is added, forming a new connected component, thus being a 0-cycle. A homology class dies when two points belonging to different connected components, i.e., they belong to two different 0-cycles, are connected by a 1-chain, thus becoming a boundary. As an example, consider the filtered complex of Figure 11.1 (top): one 0-homology class originates at K^0, two other homology classes arise at K^1; at K^2, a new homology class is created while one of the classes originated at K^1 dies, since it is merged to the class born at K^0; at K^3 another class disappears, and the same happens at K^4, where we are left with just one class that survives forever. As for 1-homology, a homology class is born when a 1-chain is added, forming a 1-cycle (for instance, a 1-simplex is added, completing a circle), while it dies when a 2-chain is added so that the 1-cycle becomes a boundary (for instance, a 2-simplex fills a circle). In the example of Figure 11.1 (top), a homology class is born at K^3, another one at K^4, then at K^5 the homology class born at K^3 dies and at K^6 also the homology class born at K^4 dies, so that no 1-cycle survives any longer. The argument goes on similarly for higher degree homology. Persistent homology algebraically captures this process of the birth and death of homology classes.

More formally, given a filtered simplicial complex $\{K^i\}_{i=0,\ldots,n}$, the *j - persistent k -th homology group* of K^i can be defined as a group isomorphic to the image of the homomorphism $\eta_k^{i,j} : H_k(K^i) \rightarrow H_k(K^{i+j})$ induced by the inclusion of K^i into K^{i+j}. Also, the *j -persistent homology group* of K^i counts how many homology classes of K^i still survive in K^{i+j}. Persistence represents the life-time of cycles in the growing filtration.

The persistent homology of a filtered complex can be represented by a set of intervals, called persistence intervals (briefly $\mathcal{P}$-*intervals*), as in Figure 11.1 (bottom): a $\mathcal{P}$-interval is a pair (i, j), with $i, j \in \mathbb{Z} \cup \{+\infty\}$ and $0 \leq i < j$, such that there exists a cycle that is completed at level i of the filtration and becomes a boundary at level j. More recently, $\mathcal{P}$-intervals have been described as sets of points in the extended planes, called *persistence diagrams*, as detailed in the next section.

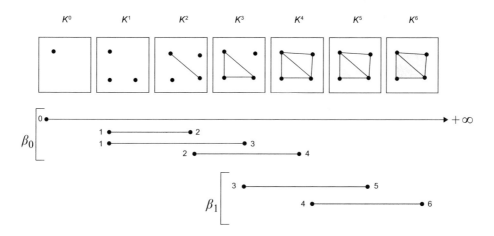

Figure 11.1: The persistent homology of a filtered complex can be represented by $\mathcal{P}$-intervals [24].

11.2   PERSISTENCE DIAGRAMS

In this section we focus on the filtration of a space X by the increasing values of a real function f defined on it. As we have seen in chapters 8–10, this is a common scenario for shape analysis applications. The function f can be used to define a filtration, made of the subspaces $X_u = f^{-1}(-\infty, u])$. Each subspace X_u includes the points of X where the function takes values less than u. These subspaces can be nested by inclusion: whenever $u < v \in \mathbb{R}$, there is an inclusion $X_u \rightarrow X_v$. According to the theory seen in the previous section, the inclusion of X_u into X_v induces a homomorphism of homology groups $H_k(X_u) \rightarrow H_k(X_v)$ for every $k \in \mathbb{Z}$, whose image is the *kth persistent homology group* of (X, f) at (u, v). The group consists of the k-homology classes that live at least from $H_k(X_u)$ to $H_k(X_v)$. Assuming that this group is finitely generated, we call its rank the *kth persistent Betti number of the pair* (X, f), and denote it by $\beta_f(u, v)$. Roughly speaking, the k-th persistent Betti number $\beta_f(u, v)$ counts the number of k-homology classes which survive while passing from X_u to X_v.

The success of persistent homology in shape analysis applications is due to the fact that a simple and compact description of the kth persistent Betti numbers of (X, f) exists, provided by the corresponding persistence diagram. Persistence diagrams are multi-sets of points in the half-plane $\Delta^+ = \{(u, v) \in \mathbb{R} \times \mathbb{R} : u < v\}$. In analogy with persistence intervals, a point (u, v) in the persistence diagram indicates that there exists a topological event that starts at level u of the filtration and ends at level v of the filtration (a cycle that is completed at level u and becomes a boundary at level v). As the length of persistence intervals represented the lifespan of topological features, in persistence diagrams the distance of a point from the diagonal $\Delta : u = v$ represents the *lifespan* of the associated topological feature. In turn, the lifespan of a feature reflects its importance: points far from the diagonal describe important or global features, i.e., the long-lived ones, whereas points close to the diagonal describe local information such as smaller details and noise.

Figure 11.2(b) shows the 0th persistence diagram obtained from the 0th persistent Betti numbers of a surface model filtered by the height function (Fig. 11.2(a)). Note that, in the case of the 0-degree homology, persistence diagrams substantially coincide with the representation of size functions by cornerpoints and cornerlines [24].

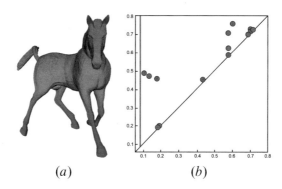

(*a*)   (*b*)

Figure 11.2: (*a*) A 3D model with the "height" function color-coded (left), and the corresponding 0th persistence diagram. The 0th persistent diagram analyzes the behavior of the connected components. The topological events registered by the red points are the birth of connected components and their death when the components joins together. The vertical line can be seen as a point at infinity, representing a topological feature that *will never die*. (*b*).

Persistence diagrams inherit their invariance properties directly from the associated function f, which can be chosen according to the application. Moreover, this kind of representation allows one to turn the problem of comparing two shapes into the simpler comparison of two corresponding persistence diagrams: this can be done by using, e.g., the *bottleneck* or the Hausdorff distances (see section 6.2), which are proven to be stable with respect to small perturbations of the data [67].

The formal definition of persistence diagrams is based on the notion of *multiplicity* [67, 83]. In what follows, the symbol Δ^* denotes the set $\Delta^+ \cup \{(u, \infty) : u \in \mathbb{R}\}$.

Definition 11.1 Multiplicity. The *multiplicity* $\mu_f(u, v)$ of $(u, v) \in \Delta^+$ is the finite, non-negative number given by

$$\min_{\substack{\varepsilon > 0 \\ u + \varepsilon < v - \varepsilon}} \quad \begin{aligned} &\beta_f(u + \varepsilon, v - \varepsilon) - \beta_f(u - \varepsilon, v - \varepsilon) + \\ &+ \beta_f(u - \varepsilon, v + \varepsilon) - \beta_f(u + \varepsilon, v + \varepsilon). \end{aligned}$$

The *multiplicity* $\mu_f(u, \infty)$ of (u, ∞) is the finite, non-negative number given by

$$\min_{\varepsilon > 0, \, u + \varepsilon < v} \beta_f(u + \varepsilon, v) - \beta_f(u - \varepsilon, v).$$

Definition 11.2 Persistence diagram. The persistence diagram $\mathrm{Dgm}(f)$ is the multiset of all points $p \in \Delta^*$ such that $\mu_f(p) > 0$, considered with their multiplicity, union the points of Δ, considered with infinite multiplicity.

The computation of persistence diagrams of filtered simplicial complexes presented in [216] exploits parallels between algebraic relationships and matrix representations. The literature also offers several techniques to speed up homology computation [88], which can be employed in the context of persistence. They generally focus on the reduction of the size of the input complex using combinatorial operations which preserve the homology.

11.3 PERSISTENCE SPACES

A recent advance in persistent homology theory is the generalization to filtrations obtained through a multi-variate function taking value in $\mathbb{R}^n$, $n > 1$. Indeed, it is often the case that the data under study (e.g., scientific data from physical or medical studies) are characterized by a large number of measurements, which can be modeled as multi-variate functions.

When f is vector-valued, i.e., $f = (f_i) : X \to \mathbb{R}^n$, the definition of the multidimensional analogue of persistent homology groups and Betti numbers is straightforward. For $u = (u_i)$, $v = (v_i) \in \mathbb{R}^n$, we say that $u \prec v$ (resp. $u \preceq v$, $u \succ v$) iff $u_i < v_i$ (resp. $u_i \leq v_i$, $u_i > v_i$) for every $i = 1, \ldots, n$. Given $u \prec v$, the *multidimensional kth persistent homology group* of (X, f) at (u, v) is the image of the homomorphism $H_k(X_u) \to H_k(X_v)$ induced in homology by the inclusion of $H_k(X_u)$ into $H_k(X_v)$. Its rank, still denoted by $\beta_f(u, v)$, is called a *multidimensional persistent Betti number*.

The multidimensional counterpart of the Definitions 11.1 and 11.2 are as follows [43]. For every $(u, v) \in \Delta_n^+ = \{(u, v) \in \mathbb{R}^n \times \mathbb{R}^n : u \prec v\}$ and $e \in \mathbb{R}^n$ with $e \succ 0$ and $u + e \prec v - e$, we consider the number

$$\begin{aligned} \mu_f^e(u, v) = &\beta_f(u + e, v - e) - \beta_f(u - e, v - e) + \\ &+ \beta_f(u - e, v + e) - \beta_f(u + e, v + e). \end{aligned} \tag{11.1}$$

Definition 11.3   Multiplicity.   The *multiplicity* $\mu_f(u, v)$ of $(u, v) \in \Delta_n^+$ is the finite, non-negative number defined by:

$$\mu_f(u, v) = \min_{\substack{e > 0 \\ u+e \prec v-e}} \mu_f^e(u, v). \tag{11.2}$$

The *multiplicity* $\mu_f(u, \infty)$ of (u, ∞) is the finite, non-negative number given by:

$$\mu_f(u, \infty) = \min_{e > 0, u+e \prec v} \beta_f(u + e, v) - \beta_f(u - e, v). \tag{11.3}$$

Having extended the notion of multiplicity to a multidimensional setting, the definition of persistence space is completely analogous to the one of persistence diagram for a real-valued continuous function.

Set $\Delta_n^* = \Delta_n^+ \cup \{(u, \infty) : u \in \mathbb{R}^n\}$ and $\Delta_n = \partial \Delta_n^+$.

Definition 11.4   Persistence Space.   The persistence space $\mathrm{Spc}(f)$ is the multiset of all points $p \in \Delta_n^*$ such that $\mu_f(p) > 0$, counted with their multiplicity, union the points of Δ_n, counted with infinite multiplicity.

Similarly to persistence diagrams, persistence spaces share invariance properties with the associated functions. Also, they can be stably compared by using, e.g., the Hausdorff distance or a generalization of the bottleneck distance.

Note that, in practice, only an approximation of $\mathrm{Spc}(f)$ can be computed. Indeed, one of the main issues in dealing with persistence spaces is that, in contrast to persistence diagrams, they are not discrete collections of points. In [39], the authors show that, in contrast with what happens for 1-dimensional persistence, it is not possible to have a *discrete and complete* descriptor of the topological information within a multi-parameter filtration. Therefore, in practical applications we can only consider approximations of persistence spaces, that is, a sample of the points with positive multiplicity. [21, 22] present solutions to the approximation and the comparison of persistence spaces.

11.4   CONCEPTS IN ACTION

Persistent homology and statistical shape analysis applied to human jaws   We have seen how the complex problem of comparing two shapes can be cast as the comparison of persistence diagrams, which is easy and stable. This has been done in [85] to compare human jaws, which is an important task in orthodontics to monitor the effects of ongoing treatments. The authors propose to incorporate the comparison of persistence diagrams in the pipeline of the standardized landmark method traditionally employed in statistical shape analysis [64], and demonstrate that this helps distinguishing clinically relevant treatment effects.

More in detail, a collection of $N = 240$ jaw bones is represented by $k = 22$ landmark points chosen by an expert for their clinical relevance. After aligning the jaw bones in $\mathbb{R}^3$, the so-called *mean shape* is computed, that is, a set of k points obtained by averaging the landmarks over the N bones. These points are used to build an abstract simplicial complex through a Delaunay triangulation.

For each of the N jaws, define the weight of an edge connecting two points as the ratio between the Euclidean distance between the landmarks corresponding to those points, and the average over all subjects of the Euclidean distance between those landmarks. The weights produce a Rips filtration (cf. section 11.1) of the Delaunay triangulation: all the nodes join at time zero, the edge between two points joins at a time corresponding to the edge weight, and higher-dimensional simplices join once all of their faces have. When this filtration is performed on an individual landmark configuration, the edges that join the filtration first are those that are smallest in that subject, relative to the entire group.

This process produces N filtrations, and, corresponding, N persistence diagrams, one for each jaw. Actually, $3N$ persistence diagrams are computed, namely the 0th, 1st, and 2nd persistence diagram. The Wasserstein distance between persistence diagrams is computed to obtain four $N \times N$ matrices of pairwise distances between pairs of jaws: one $N \times N$ for each dimension (0, 1, 2), and a cumulative matrix. Multidimensional scaling of the matrices is performed to embed the data as points in $\mathbb{R}^2$.

A close analysis of these data reveals a correlation between one of the embedding coordinates and the treatment used on the patients (jaw expansion), enabling one to distinguish between the control group and the two treatment groups as the treatment evolves over time. This demonstrates that the persistent homology method is able to distinguish clinically relevant treatment effects.

Another example of application of persistence diagrams and spaces for textured 3D shape retrieval will be presented in chapter 12.

CHAPTER 12

Beyond Geometry and Topology

Throughout this book we have focused on mathematical tools for shape analysis, where the term *shape* mainly referred to geometrical and topological properties of objects. Nonetheless, there is much more to deal with the problem of 3D shape analysis.

A first issue concerns appearance properties, coded by *texture*. Nowadays 3D textured models can be easily obtained, as most sensors are able to acquire both 3D shape and texture with reasonable accuracy, and multiple-view stereo techniques enable the recovery of both geometric and colorimetric information directly from images. Examples of textured 3D models can be seen in Figure 12.1. Texture and colorimetric features contain rich information about the visual appearance of real objects. For example, in 3D cultural heritage the visual appearance of objects (materials, decorations, etc.) conveys essential historical and artistic information. It enables the categorization of objects according to surface decorations, or to attribute the fabric of furnishings to a given producer, or even to match degradation phenomena on element surfaces (e.g., for conservation and restoration purposes). Though most of the shape analysis techniques developed so far did not take texture into account, the attention towards appearance properties has considerably grown over the last few years, and the first techniques which analyze both shape and appearance properties have been proposed [20, 121, 129, 160, 172, 199, 213].

Figure 12.1: Models from the "Retrieval of textured 3D models" track of the SHREC'13 benchmark [42].

Besides geometry and texture, *semantic* aspects are of fundamental importance to 3D shape analysis. Here *semantics* refers to the meaning, or *functionality*, of an object in a given context. A variety of recent 3D shape analysis techniques are meant to derive semantic (high-level) information from low-level properties. A first example is *semantic annotation*, which is the automatic or semi-automatic labeling of objects (or parts of objects) with a tag describing their content [7, 122]. Then, *attribute transfer* techniques study how to automatically transfer labels from a single object (or part of an object) to sets of unknown objects [120, 127]. Another example is *viewpoint selection*, referring to the process of finding the most informative view of an object in a given context [124, 146].

A recent path of research aims to derive high-level information about objects by analyzing single objects in the context of larger *collections* of objects: the idea is to derive information not only from the object itself, but also from its relation with the other objects in the collection. This is the case for example of the *co-segmentation* of a set of 3D objects [105, 108, 208], i.e., the segmentation of the objects as a whole into consistent semantic parts with part correspondences. Another example is the structuring of 3D large datasets to enable navigation and retrieval; this is often achieved by exploiting not only the similarity between a query and the dataset, but also the pairwise similarities between the rest of dataset models.

In this chapter, we show how mathematical concepts can support 3D shape analysis even beyond the analysis of geometric aspects, by taking into account texture and semantic aspects. We will start with examples of 3D textured shape retrieval (section 12.1). Then, we present a technique for the qualitative organization of datasets of 3D models (section 12.2). Finally, we illustrate a method to infer the functionality of an object through co-segmentation and co-analysis (section 12.3). Despite the focus on the technical aspects of the methods, we hope the choice of these techniques will stimulate further thoughts and investigations in the field.

12.1   3D TEXTURED SHAPE RETRIEVAL

The authors of [121] generalized heat kernels (section 5.2) to textured shapes. The main idea is to define the diffusion process, hence the heat kernel, on a manifold embedded into a high-dimensional space, instead of the usual 3-dimensional Euclidean space $\mathbb{R}^3$: namely, a 6-dimensional space $\mathbb{R}^3 \times C$, where C is some color space, for example the RGB or the CIELab color spaces. In this embedding, three coordinates represent geometric information, in terms of the usual Cartesian coordinates, whereas the other three represent the photometric information, in terms of color components. The definition of the Laplace-Beltrami operator, hence the diffusion process and the heat kernel, are straightforward. There it follows the definition of a photometric Heat Kernel Signature (pHKS), which is used for textured shape retrieval. Similar ideas have been used in [107], who defined a family of diffusion distances based on Schrödinger operators incorporating photometric data.

The authors of [23] proposed PHOG, a signature for 3D textured shape retrieval which is able to analyze both colorimetric and geometric properties. The main idea is to look at colorimetric properties as at multi-value attributes associated with the vertices of a model, and taking advantage of the persistence homology settings (chapter 11). The PHOG signature consists of three parts:

- a colorimetric descriptor: coordinates in the CIELab color space are seen as either scalar (*L* channel, i.e., luminosity) or bivariate (*a* and *b* channels, i.e., color hues) real functions defined over the shape, and used to get persistence diagrams and (approximated) persistence spaces;

- a hybrid descriptor: shape and texture are jointly captured by persistent diagrams working on a generalization of the integral geodesic distance which fuses shape and color properties and allows the user to weight them according to his/her needs. Then, the hybrid descriptor is a set of persistence diagrams of the generalized geodesic distances;

- a geometric descriptor: a stochastic matrix of distances among real functions defined on 3D objects defined on the basis of the technique described in [18]. A selection of representative functions of the objects can be done through the *DBSCAN* clustering technique [76].

For two shapes S_1 and S_2, their distance is the sum of the distances between the colorimetric, hybrid and geometric descriptors. In particular, the photometric distance is the normalized sum of the bottleneck distance between the persistence diagram of the L channel and of the Hausdorff distance between the persistence spaces of the bi-variate function defined by the a and b channels (see sections 6.2, 11.2 and 11.3 for details on distance definitions and persistence theory). Similarly, the hybrid distance is the normalized sum of the bottleneck distance between the persistence diagrams of the generalized geodesic distances. Finally, the geometric distance is computed as the Manhattan distance between the associated stochastic matrices of geometric functions.

Examples of retrieval results based on these descriptors is shown in Figure 12.2. The models are ordered from left to right; the first column represents the query model. The results in the first row correspond to the texture descriptor, the second row to the hybrid and the third row to the geometric descriptor; finally the fourth row when the combined PHOG description is considered. It is interesting to note how different features are taken into account according to the chosen descriptor.

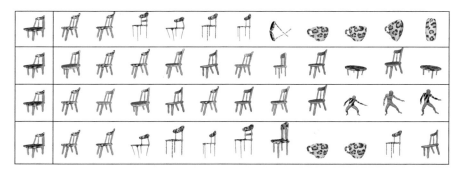

Figure 12.2: From high to below: retrieved items with respect to: CIELab, hybrid, geometric and combined descriptions [42].

12.2   QUALITATIVE ORGANIZATION OF COLLECTIONS OF 3D MODELS

An ever growing number of 3D models are produced and stored in online shape repositories (e.g., Trimble 3D warehouse, Turbosquid, see also section 13.2). This demands techniques to organize large, heterogeneous collections of 3D models, so as to understand their overall categorization and summarize their content. The goal is to facilitate exploration and content search. Any organization of these large collections must be built on a comparison method between the individual shapes, and the success of the organization highly depends on how we choose the comparison strategy. The key challenge is that shape can vary in different ways, and users may be interested in different types of variations [112, 113]. This is why a single measure is not likely to provide a good organization. Moreover, the authors in [106] observe that *quantitative* measures may be unreliable, or at least not informative enough, when the collections possess rich variation and highly dissimilar objects: it is expected that a numerical distance between a table and a car is less informative than the distance between two tables.

Therefore, the idea in [106] is to use *qualitative* information derived from multiple quantitative measures. The so-called *quartets* are sought in the dataset, which are quadruplets of 3D objects divided into two pairs. Each pair is made of two similar objects, which are clearly different from the other pair (Figure 12.3, left). Quartets incorporate several distance measures into qualitative, topological relations of closeness and separation between objects. They are used to build a so-called *categorization tree*, or *C-tree*, which organizes the collection so that 3D shapes reside at the leaves, with the number of edge hops between them reflecting their degree of separation (Figure 12.3, middle). The C-tree must maintain the topological relations defined by quartets. Given a 3D object S, the C-tree is used to partition all other objects in the collection into layers, organized by their degree of separation around S (Figure 12.3, right). This type of organization is called the *Degree of Separation (DoS) chart*. The DoS enables the interactive exploration of shape collections through a 2D display space. Once the user selects a query object among the leaves of the C-tree, the rest of the objects are automatically repositioned to form a DoS chart around the query: shapes close to the query in the C-tree will be located at the inner circles of the DoS chart, thus giving the user an intuitive understanding of how the objects in the collection match with the query.

From a technical standpoint, the main issues are defining reliable quartets and building the C-tree out of them. The representation in Figure 12.4(a) shows a quadruplet of objects (a, b, c, d) as the vertices of a square, with the six edges representing the distances between them, according to different similarity metrics. A quadruplet is a candidate quartet if its four vertices remain connected after removing the three edges corresponding to the largest distances; otherwise it is discarded (Figure 12.4(e)). Then, the edge corresponding to the largest distance must be a bridge, that is, its removal separates the four nodes into two pairs; otherwise, the quadruplet is discarded again (Figure 12.4(d)). Those candidate quartets (Figure 12.4(b) and (c)) are further filtered by imposing a threshold on the ratio of the remaining three edges, to ensure that there are well separated pairs made of enough close objects. Four shape descriptors are used along with their distances, namely the LightField descriptor [44], the Spherical Harmonics descriptor [84], the Bag-of-Words based on the Heat Kernel Signature [37], the Shape Diameter Function [179]. A set of reliable quartets is created for each descriptor separately, then they are combined to build the C-tree.

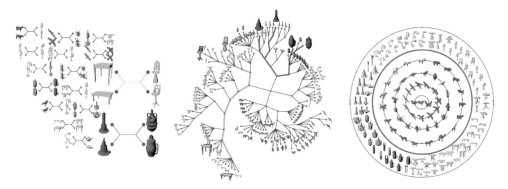

Figure 12.3: A collection of 3D objects is organized in a categorization tree (middle) and a Degree of Separation chart (right) via quartet-based analysis (left) [106].

The C-tree is built from the quartets using a two-stage algorithm: first embedding, then partitioning. Shapes are represented as points in the unit sphere S^n. An energy function is defined as the difference between two terms: the sum over quartets of the angles between close objects, and the sum over quartets of the angles between distant objects. Minimizing this energy function yields optimized points which respect the topological relations imposed by the quartets. Finally, also the partitioning is guided by information from the quartets. Naming *bad* edges those between close objects and *good* edges those between distant objects, the partition is sought which maximizes the number of cuts through good edges while minimizing the number of cuts through bad ones. This is done using the Quartets MaxCut (QMC) algorithm [186]. The partition is operated recursively until no more partitions are obtained. The resulting hierarchy of sub-partitions defines the C-tree.

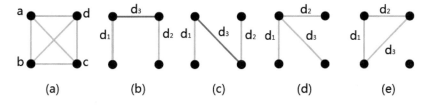

Figure 12.4: Four objects and six associated edges (a). Candidate quartets (b and c) and quadruplets to be discarded (d and e) [106].

12.3 RECOGNITION OF FUNCTIONAL PARTS OF MAN-MADE OBJECTS

Laga et al. in [122] address the problem of automatic recognition of functional parts of man-made 3D shapes, starting from classes of objects that present significant geometric and topological variations. The idea is to infer the context of a part within a 3D shape and to use it to learn the semantics of shapes. In practice, the context is modeled in terms of structural relationships between shape parts that are used, in addition to part geometry, as cues for functionality recognition.

Each shape is represented as a graph that connects the parts that share some spatial relationships and the context of a shape part is modeled as walks in the graph, adopting a graph kernel strategy. Each node of the graph corresponds to a shape part and two nodes are connected with an edge if they share some contextual relationship. During the initial graph construction, two types of relationships are considered: (i) *inter-part symmetries* if there exists a rotational, translational, or reflectional symmetry transform that aligns two parts; (ii) *adjacency* if two parts share some vertices or if their bounding boxes overlap. Then, a hierarchical graph is created, by contracting the cliques of the original graph on the basis of their geometry and adjacency. Nodes within the same level are connected with either an adjacency or a symmetry relationship and a node is connected to its parent with an edge.

Geometric attributes are assigned to each node to encode the geometric properties of its shape and provide a mean for estimating the similarity between the geometry of two nodes. In this method the geometric properties chosen are the distance-based shape distribution proposed in [152] (shape descriptor), the radius of the sphere that bounds the component (size) and the three eigenvalues of the Principal Component Analysis (PCA) of each shape part (component aspect). Then to compare the geometry of two nodes, a Gaussian kernel for each of the three geometric properties is built and the geometric kernel is obtained as the uniformly weighted sum of the three components.

The contextual relationships are the core component to the proposed shape correspondence and functionality learning algorithms. The basic assumption is that components sharing common contextual relations are more likely to have the same functionality. The relations of interest include enclosure, side contact, symmetry, co-centricity, and adjacency (see Figure 12.5 left and middle). To define a similarity metric between two edges labeled with the relationship types it is assumed that the two edges are similar if they are of the same type. As the final assumption, two parts are similar if their geometry and context are similar. While geometry is informative, under large geometric and topological variations, the context, i.e., the way shape parts are interconnected, provides rich information about the semantics of shapes. As context-aware similarity measure between two nodes, all walks of length p starting from the two nodes are considered; in the practice, the parameter p is set with the value 3.

Finally, the functional recognition is seen as a multi-label supervised classification problem that assigns a shape part to one of the possible functional categories and it is solved with a Support

Figure 12.5: Illustration of the flow of how mathematical abstraction is used to extract semantically significant parts of a shape (or a template) and how to co-analyze a set of 3D objects belonging to the same class.

Vector Machine (SVM) technique, in this specific case the implementation is provided in WEKA software.[1]

[1]http://www.cs.waikato.ac.nz/ml/weka/

CHAPTER 13

Resources

We conclude our book around shape analysis with a set of pointers to resources used in the world of shape analysis. The choice of items has been based on the authors' experience as researchers in the field. Though the list is not exhaustive, it is meant as a starting point to investigate the literature on 3D shape analysis. The list below includes software, repositories of 3D models, and benchmarks. Additionally, the bibliography at the end of the book references prominent conferences, journals, and books, which should offer the reader high-quality information on 3D shape analysis problems and methods. Besides those cited below, we provide a catalogue of further resources at web site [2], where readers may find additional pointers to infrastructures, events and projects on the topics of this book.

13.1 SOFTWARE

The software tools selected include both geometry processing tools (e.g., shape reconstruction, mesh editing) and methods for high-level data analysis (e.g., shape segmentation, similarity assessment).

- Computational Geometry Algorithms Library (CGAL): `https://www.cgal.org/`. CGAL offers data structures and algorithms for model representation, arrangements of curves, mesh generation, geometry processing, and shape analysis, fitting, and distance evaluation. CGAL is used in various areas needing geometric computation, such as: computer graphics, scientific visualization, computer aided design and modeling, geographic information systems, molecular biology, medical imaging, robotics and motion planning, mesh generation, numerical methods, etc.

- Point Cloud Library (PCL): `http://pointclouds.org/`. The PCL library is a standalone, large scale, open project for 2D/3D image and point cloud processing. It is a bit far from the focus of the book but it is largely used to pre-process point clouds data.

- Eigen: `http://eigen.tuxfamily.org/index.php?title=Main_Page`. Eigen is a C++ template library for linear algebra: matrices, vectors, numerical solvers, and related algorithms. It is not a strictly a 3D shape analysis library but it is largely used in several tools for 3D shape analysis.

- ViennaCL: `http://viennacl.sourceforge.net/`. Similarly to Eigen, this resource provides a linear algebra library that can be used for several 3D analysis tools. In particular, this library supports computations on many-core architectures (GPUs, MIC) and multi-core CPUs.

• MeshLab: http://meshlab.sourceforge.net/. This software provides a system for processing and editing unstructured 3D triangular meshes rather than a 3D analysis system. MeshLab is a useful tool to pre-process unstructured models arising in 3D scanning, providing a set of tools for editing, cleaning, healing, inspecting, rendering and converting these meshes.

• The Visualization and Computer Graphics Library (VCG): http://vcg.isti.cnr.it/~cignoni/newvcglib/html/. The VCG is a library for manipulation, processing and displaying with OpenGL of both triangle and tetrahedral meshes. In particular, the library implements edge-collapse based simplification, smoothing and fairing algorithms, curvature estimation, Hausdorff distance computation; it also extracts geodesic paths and isosurfaces and generates subdivision surfaces.

• ReMesh (JMeshLib): http://remesh.sourceforge.net/. Similarly to MeshLab, ReMesh offers an editing tool for manifold triangle meshes with repairing features. In particular, ReMesh maybe complemented with MexFi: http://meshfix.sourceforge.net/ a tool able to convert raw digitized polygon meshes to a clean model where holes, self-intersections, degenerate and non-manifold elements are replaced with valid configurations.

• OpenMesh: http://www.openmesh.org/. OpenMesh is a data structure for representing and manipulating polygonal meshes, not only triangle meshes that can be included in more sophisticated tools for 3D object representation and manipulation. For instance, OpenFlipper: http://www.openflipper.org/, is an example of geometry modeling and processing framework developed on the top of OpenMesh.

• Toolbox graph: http://www.mathworks.com/matlabcentral/fileexchange/5355-toolbox-graph. This toolbox contains useful functions to deal with graphs and triangulations, in particular it allows the computation of mean and Gaussian curvatures and offers several tools originated from spectral theory (Laplace-Beltrami operators, diffusion kernels, etc.).

• MeshLP: Approximating Laplace-Beltrami Operator from Meshes, http://www.geomtop.org/software/meshlp.html. MeshLP offers an implementation of the mesh Laplace operator which approximates the Laplace-Beltrami operator of the surface approximated by the input triangle mesh. On the top of MeshLP, the Heat Kernel Signature (HKS) is publicly available at http://www.geomtop.org/software/hks.html (see description of the method in Sect. 5.2).

• Sparse ICP: http://lgg.epfl.ch/sparseicp is a tool for rigid registration of two geometric data sets represented as point clouds.

- Shape characterization and structuring through Reeb graph Computation (ShReC): `http://saturno.ge.imati.cnr.it/ima/smg/shrec-web/SHREC.htm`. SHREC is a tool for characterizing (in the sense of computing maxima, minima and saddles) a surface with or without boundary with respect to a real function by computing a set of contour levels. Starting from such a characterization, SHREC computes an Extended Reeb graph representation of the surface, according to the method described in [6] and [17].

- Multi-Scale Segmentation of Triangulated Surfaces: `http://saturno.ge.imati.cnr.it/ima/smg/plumber-web/plumber-web.html`. This tool implements the multi-scale curvature characterization of a closed surface described in section 4.2 and automatically extracts the features that can be described as generalized cylinders or cones.

- SI-HKS & ISC: `http://vision.mas.ecp.fr/Personnel/iasonas/descriptors.html`. These tools are freely available examples of non-rigid shape descriptions that have been applied to similarity analysis of images and 3D shapes.

- TransforMesh, MeshHOG and MVViewer: `http://mvviewer.gforge.inria.fr/`. This repository contains an implementation of a 3D feature detector (MeshDOG) and a 3D feature descriptor (MeshHOG) for uniformly triangulated meshes, invariant to changes in rotation, translation, and scale. Both the detector and the descriptor are able to capture the local geometric and/or photometric properties.

13.2 3D DATASETS AND BENCHMARKS

13.2.1 3D DATASETS

In the following we list a selection of 3D model repositories that are available on the web and used by the computer graphics community as sources of data for 3D shape analysis.

- Trimble 3D warehouse: `https://3dwarehouse.sketchup.com/` formerly Google 3D Warehouse `http://sketchup.google.com/3dwarehouse/`

- 3DVia: `http://www.3dvia.com/products/3dvia-shape/`

- 3D repository: `http://3dr.adlnet.gov`

- Turbosquid: `http://www.turbosquid.com/3d`

- Tosca: `http://tosca.cs.technion.ac.il/book/resources{_}data.html`

- 3Dcafe: `http://www.3dcafe.com/free-3d-models`

- The Shape repository of the "AIMSHAPE" project: `http://shapes.aim-at-shape.net/`

- The Shape repository of the Visualization Virtual Services (VVs) "VISIONAIR:" `http://visionair.ge.imati.cnr.it/ontologies/shapes/`, which also offers services for preparing digital shapes for visualization purposes, and shape search facilities.

13.2.2 BENCHMARKS AND CONTESTS

Enabling the reproduction and evaluation of algorithms is an important issue in computer science. Several benchmarking initiatives exist in the multimedia domain, such as TRECVID [153, 184, 185] and IMAGECLEF [148]. The computer graphics community recognized opportunity in benchmarking as a means to ensure reproducibility of results: indeed, with the ever-growing number of techniques proposed, it is imperative to provide users with tools to decide on which solution is best suited to the application at hand. The creation of standard datasets with a ground truth and a "quality label" paves the road to a fair evaluation of the existing technologies, as well as the identification of new directions of research. Therefore, benchmarks were proposed for example for surface reconstruction from point clouds [16], 3D retrieval (SHREC [205]), and keypoint detection [200].

In 2011, a world-leading scientific publisher, Elsevier, launched the *Executable Papers Grand Challenge*, pushing the idea of benchmarking towards new forms of publication. The goal was to address the question of how to reproduce computational results within the confines of the research article. The winner, the Collage Authoring Environment, launched a pilot special issue on 3D Object Retrieval with the journal *Computers & Graphics*, showcasing executable research results in articles [4]. In an executable paper, authors can embed chunks of executable code and data into their papers, and readers may execute that code within the framework of the research article.

In the following, we focus on benchmarking initiatives addressing two topics, namely shape retrieval and segmentation, which are very relevant to the shape analysis, and provide a substantial number of interesting 3D shapes to be used for experimenting with shape analysis techniques.

Retrieval Among the others, the Shape Retrieval Contest (SHREC) (`http://www.aimatshape.net/event/SHREC/`) was proposed with the general objective of evaluating the performances of 3D shape matching and retrieval algorithms [205]. The initial results of the contest provided the first opportunity to analyze a number of state-of-the-art algorithms, their strengths, as well as their weaknesses, using a common test collection allowing for a direct comparison of algorithms. SHREC provides many resources to compare and evaluate 3D retrieval methods, in particular ground truths and statistical measures, such as precision recall or first and second tier. Started in 2006, SHREC is still ongoing. Indeed, a single test collection necessarily delivers only a partial view of the whole picture, hence the contest quickly moved towards a multi-track organization, with specific tracks for the single aspects of the problem (e.g., global and partial matching) [19, 36, 38, 42], with a distinction among the query representations (e.g., polygon soup, sketches and watertight models), as well as a number of context-specific benchmarks, for example for mechanical part matching, molecule matching, or 3D face matching.

As an example, in 2014, five tracks were organized:

1. Automatic Location of Landmarks used in Manual Anthropometry (`http://www.andrea giachetti.it/shrec14/`);

2. Shape Retrieval of Non-Rigid 3D Human Models (`http://www.cs.cf.ac.uk/shaper etrieval/shrec14/index.html`);

3. Retrieval and Classification on Textured 3D Models (`http://www.ge.imati.cnr.it/sh rec14`);

4. Extended Large Scale Sketch-Based 3D Shape Retrieval (`http://www.itl.nist.gov/i ad/vug/sharp/contest/2014/SBR/`);

5. Large Scale Comprehensive 3D Shape Retrieval (`http://www.itl.nist.gov/iad/vug /sharp/contest/2014/Generic3D/`).

Segmentation   The mesh segmentation benchmark `http://segeval.cs.princeton.edu/` provides 380 meshes across 19 object categories for quantitative analysis of how people decompose objects into parts and for comparison of automatic mesh segmentation algorithms. To build the benchmark, eighty people were recruited to manually segment surface meshes into functional parts, yielding an average of 11 human-generated segmentations for each mesh. This data set provides a sampled distribution over "how humans decompose each mesh into functional parts," and this knowledge is treated as a probabilistic "ground truth." In addition, the dataset is coupled with software for evaluation, analysis, and viewing of mesh segmentations. The code is written in C++ and free to use. Also Python scripts to automatically run evaluation and analysis experiments are available to plot the results (using Matlab), to create reports, as well as to generate colored images of mesh segmentations. The idea is that this benchmark can be used to study and compare new segmentation algorithms, provided that the results of seven methods are given together with the human-generated mesh segmentations. The survey [45] demonstrates that there is not a single algorithm that performs well on all shape classes.

Another shape segmentation benchmark is the 3D Segmentation Benchmark [15] available at `http://www-rech.telecom-lille1.eu:8080/3dsegbenchmark/`. Similarly to [45], this benchmark provides twenty-eight 3D-models grouped in five classes, namely animal, furniture, hand, human and bust. Each 3D-model is associated with some manual segmentations done by volunteers. A small number of varied models with respect to a set of properties is selected. All the selected models are manifold, connected, and do not have intersecting faces. Hence they are supported as an input by any segmentation algorithm. In order to collect precise manual segmentations, the volunteers have been assisted when traced the vertex-boundaries through the different models even if no condition was imposed on the manner with which they have segmented them.

To evaluate a new segmentation algorithm it is necessary to create a user account, upload the files which are the results of the segmentation algorithm and the evaluation result are displayed online, in a temporary page. In addition, the results of eight segmentation algorithms are available and can be visualized with the new method and exported under different formats (PDF, CVS, etc.).

Finally, also a benchmark for co-segmentation (the so-called Shape COSEG Dataset) [182] is available at `http://web.siat.ac.cn/~yunhai/ssl/ssd.htm` (see Figure 13.1). The goal of this work is to provide data for quantitative analysis of how people consistently segment a set of shapes and for evaluation of our active co-analysis algorithm. The dataset is a collection of eleven classes of shapes which possess a consistent ground-truth segmentation and labeling. Among them, seven sets come from the dataset of Sidi et al. [2011] while a small, but challenging, class of iron models has been created. In addition, to consider the labeling of large sets, three additional large sets have been created: tele-aliens, vases and chairs. These three sets consists of 200, 300, 400 shapes, respectively.

Figure 13.1: The COSEG main page.

In addition to the benchmarks here described, other references must be cited: Princeton Shape Benchmark: `http://shape.cs.princeton.edu/benchmark/` and the McGill 3D Shape Benchmark: `http://www.cim.mcgill.ca/~shape/benchMark/`.

Bibliography

[1] Discrete differential geometry forum at the columbia university. `http://ddg.cs.colum bia.edu/`. 5

[2] Mathematical tools for shape analysis and description. `http://www.ge.imati.cnr.it/ mathematical_tools`. 97

[3] *Longman Dictionary of Contemporary English*. Longman, 1978. 2

[4] Executable paper special issue. `http://www.journals.elsevier.com/computers-an d-graphics/news/introducing-exe%cutable-papers/`, 2013. 100

[5] V. I. Arnold. *Lectures on partial differential equations*. Universitext (1979). Springer, 2004. 28

[6] M. Attene, S. Biasotti, and M. Spagnuolo. Shape understanding by contour-driven retiling. *The Visual Computer*, 19(2-3):127–138, 2003. 70, 72, 99

[7] M. Attene, F. Robbiano, M. Spagnuolo, and B. Falcidieno. Characterization of 3D shape parts for semantic annotation. *Computer Aided Design*, 41(10):756–763, October 2009. DOI: 10.1016/j.cad.2009.01.003. 89

[8] A. J. Baddeley. Errors in binary images and an Lp version of the Hausdorff metric. *Nieuw Arch. Wiskunde*, 10:157–183, 1992. 49

[9] C. L. Bajaj, V. Pascucci, and D. R. Schikore. The contour spectrum. In *VIS '97: Proceedings of the IEEE Visualization 1997*, pages 167–173. IEEE Computer Society Press, 1997. DOI: 10.1109/VISUAL.1997.663875. 63

[10] T. F. Banchoff. Critical points and curvature for embedded polyhedra. *Journal of Differential Geometry*, 1:245–256, 1967. DOI: 10.2307/2317380. 63

[11] T. F. Banchoff. Critical points and curvature for embedded polyhedral surfaces. *American Mathematical Monthly*, 77:475–485, 1970. DOI: 10.2307/2317380. 63

[12] V. Barra and S. Biasotti. 3D shape retrieval using kernels on Extended Reeb graphs. *Pattern Recognition*, 46(11):2985 – 2999, 2013. DOI: 10.1016/j.patcog.2013.03.019. 72, 73

[13] V. Barra and S. Biasotti. Learning kernels on Extended Reeb graphs for 3D shape classification and retrieval. In *Proceedings of the Sixth Eurographics Workshop on 3D Object Retrieval*, 3DOR '13, pages 25–32, Aire-la-Ville, Switzerland, Switzerland, 2013. Eurographics Association. DOI: 10.1007/s00371-014-0926-5. 74

[14] M. Belkin, J. Sun, and Y. Wang. Discrete Laplace operator for meshed surfaces. In *Symposium on Computational Geometry*, 2008. DOI: 10.1145/1377676.1377725. 37

[15] H. Benhabiles, J.-P. Vandeborre, G. Lavoué, and M. Daoudi. A framework for the objective evaluation of segmentation algorithms using a ground-truth of human segmented 3D-models. In *IEEE International Conference on Shape Modeling and Applications (Shape Modeling International 2009)*, Beijing, China, June 26–28 2009. short paper. DOI: 10.1109/SMI.2009.5170161. 101

[16] M. Berger, J. A. Levine, L. G. Nonato, G. Taubin, and C. T. Silva. A benchmark for surface reconstruction. *ACM Transactions on Graphics*, 32(2):20:1–20:17, 2013. DOI: 10.1145/2451236.2451246. 100

[17] S. Biasotti. Topological coding of surfaces with boundary using Reeb graphs. *Computer Graphics and Geometry*, 7(1):31–45, 2005. 99

[18] S. Biasotti. Shape comparison through mutual distances of real functions. In *Proceedings of the ACM workshop on 3D object retrieval*, 3DOR '10, pages 33–38, New York, NY, USA, 2010. ACM. DOI: 10.1145/1877808.1877816. 91

[19] S. Biasotti, X. Bai, B. Bustos, A. Cerri, D. Giorgi, L. Li, M. Mortara, I. Sipiran, S. Zhang, and M. Spagnuolo. Shrec'12 track: Stability on abstract shapes. In M. Spagnuolo, M. Bronstein, A. Bronstein, and A. Ferreira, editors, *3DOR*, pages 101–107, Cagliari, Italy, 2012. 100

[20] S. Biasotti, A. Cerri, M. Abdelrahman, M. Aono, A. Ben Hamza, M. El-Melegy, A. Farag, V. Garro, A. Giachetti, D. Giorgi, A. Godil, C. Li, Y.-J. Liu, H. Y. Martono, C. Sanada, A. Tatsuma, S. Velasco-Forero, and C.-X. Xu. Shrec 2014 track: Retrieval and Classification on Textured 3D Models. In B. Bustos, H. Tabia, J.-P. Vandeborre, and R. Veltkamp, editors, *3DOR*, pages 111–120, 2014. 89

[21] S. Biasotti, A. Cerri, P. Frosini, and D. Giorgi. A new algorithm for computing the 2-dimensional matching distance between size functions. *Pattern Recognition Letters*, 7(32):1735–1746, 2011. DOI: 10.1016/j.patrec.2011.07.014. 86

[22] S. Biasotti, A. Cerri, P. Frosini, D. Giorgi, and Landi C. Multidimensional size functions for shape comparison. *Journal of Mathematical Imaging and Vision*, 7(32):161–179, 2008. DOI: 10.1007/s10851-008-0096-z. 86

[23] S. Biasotti, A. Cerri, D. Giorgi, and M. Spagnuolo. PHOG: Photometric and Geometric Functions for Textured Shape Retrieval. *Computer Graphics Forum*, 32(5):13–22, 2013. DOI: 10.1111/cgf.12168. 90

[24] S. Biasotti, L. De Floriani, B. Falcidieno, P. Frosini, D. Giorgi, C. Landi, L. Papaleo, and M. Spagnuolo. Describing shapes by geometrical-topological properties of real functions. *ACM Computing Surveys*, 40(4):1–87, 2008. DOI: 10.1145/1391729.1391731. 54, 55, 57, 62, 63, 66, 71, 83, 84

[25] S. Biasotti, D. Giorgi, M. Spagnuolo, and B. Falcidieno. Reeb graphs for shape analysis and applications. *Theoretical Computer Science*, 392(1–3):5–22, 2008. DOI: 10.1016/j.tcs.2007.10.018. 72

[26] S. Biasotti, M. Marini, M. Spagnuolo, and B. Falcidieno. Sub-part correspondence by structural descriptors of 3D shapes. *Computer Aided Design*, 38(9):1002–1019, September 2006. DOI: 10.1016/j.cad.2006.07.003. 74

[27] A. I. Bobenko and B. A. Springborn. A discrete Laplace-Beltrami operator for simplicial surfaces. *Discrete & Computational Geometry*, 38(4):740–756, December 2007. DOI: 10.1007/s00454-007-9006-1. 37

[28] B. Bonev, F. Escolano, D. Giorgi, and S. Biasotti. Information-theoretic selection of high-dimensional spectral features for structural recognition. *Computer Vision and Image Understanding*, 117(3):214–228, 2013. DOI: 10.1016/j.cviu.2012.11.007. 72

[29] P. Bose, A. Maheshwari, C. Shu, and S. Wuhrer. A survey of geodesic paths on 3D surfaces. *Computational Geometry*, 44(9):486 – 498, 2011. DOI: 10.1016/j.comgeo.2011.05.006. 27, 28

[30] R. L. Boyell and H. Ruston. Hybrid techniques for real-time radar simulation. In *Proceedings of the 1963 Fall Joint Computer Conference*, Nov 1963. DOI: 10.1145/1463822.1463869. 70

[31] P.-T. Bremer, H. Edelsbrunner, B. Hamann, and V. Pascucci. A multi-resolution data structure for two-dimensional Morse functions. In G. Turk, J. van Wijk, and R. Moorhead, editors, *VIS '03: Proceedings of the IEEE Visualization 2003*, pages 139–146. IEEE Computer Society Press, October 2003. 77

[32] P.-T. Bremer, H. Edelsbrunner, B. Hamann, and V. Pascucci. A topological hierarchy for functions on triangulated surfaces. *IEEE Transactions on Visualization and Computer Graphics*, 10(4):385–396, July/August 2004. DOI: 10.1109/TVCG.2004.3. 78

[33] P. T. Bremer and V. Pascucci. A practical approach to two-dimensional scalar topology. In H. Hauser, H. Hagen, and H. Theisel, editors, *Topology-based Methods in Visualization*,

pages 151–169. Springer Berlin Heidelberg, 2007. DOI: 10.1007/978-3-540-70823-0. xiv, 78, 79

[34] P.-T. Bremer, V. Pascucci, and B. Hamann. Maximizing adaptivity in hierarchical topological models. In *SMI '05: Proceedings of the International Conference on Shape Modeling and Applications 2005*, pages 300–309. IEEE Computer Society Press, 2005. DOI: 10.1109/SMI.2005.28. 78

[35] A. Bronstein, M. Bronstein, and R. Kimmel. *Numerical Geometry of Non-Rigid Shapes*. Springer Publishing Company, Incorporated, 1 edition, 2008. 16, 37

[36] A. M. Bronstein, M. M. Bronstein, U. Castellani, B. Falcidieno, A. Fusiello, A. Godil, L. J. Guibas, I. Kokkinos, Z. Lian, M. Ovsjanikov, G. Patané, M. Spagnuolo, and R. Toldo. SHREC 2010: Robust large-scale shape retrieval benchmark. In *Proc. EUROGRAPHICS Workshop on 3D Object Retrieval (3DOR)*, pages 71–78, 2010. 100

[37] A. M. Bronstein, M. M. Bronstein, L. J. Guibas, and M. Ovsjanikov. Shape google: Geometric words and expressions for invariant shape retrieval. *ACM Trans. Graph.*, 30(1):1:1–1:20, February 2011. DOI: 10.1145/1899404.1899405. 93

[38] A. M. Bronstein, Michael Bronstein, B. Bustos, U. Castellani, M. Crisani, B. Falcidieno, L. J. Guibas, I. Sipiran, I. Kokkinos, V. Murino, M. Ovsjanikov, G. Patané, M. Spagnuolo, and J Sun. SHREC 2010: robust feature detection and description benchmark. In *Proc. EUROGRAPHICS Workshop on 3D Object Retrieval (3DOR)*, page 87–91, 2010. DOI: 10.1145/1899404.1899405. 100

[39] G. Carlsson and A. Zomorodian. The theory of multidimensional persistence. *Discrete Comput. Geom.*, 42(1):71–93, 2009. DOI: 10.1007/s00454-009-9176-0. 86

[40] H. Carr, J. Snoeyink, and U. Axen. Computing contour trees in all dimensions. In *SODA '00: Proceedings of the 11th Annual ACM-SIAM Symposium on Discrete Algorithms*, pages 918–926, Philadelphia, PA, USA, 2000. ACM Press. DOI: 10.1016/S0925-7721(02)00093-7. 70

[41] F. Cazals, F. Chazal, and T. Lewiner. Molecular shape analysis based upon the Morse-Smale complex and the Connolly function. In *SCG '03: Proceedings of the 19 * th Annual Symposium on Computational Geometry*, pages 351–360, New York, NY, USA, 2003. ACM Press. DOI: 10.1145/777792.777845. 77

[42] A. Cerri, S. Biasotti, M. Abdelrahman, J. Angulo, K. Berger, L. Chevallier, M. T. El-Melegy, A. A. Farag, F. Lefebvre, A. Giachetti, H. Guermoud, Y.-J. Liu, S. Velasco-Forero, J.-R. Vigouroux, C.-X. Xu, and J.-B. Zhang. SHREC'13 track: Retrieval on textured 3D models. In *3DOR*, pages 73–80, 2013. DOI: 10.2312/3DOR/3DOR13/073-080. 89, 91, 100

[43] A. Cerri and C. Landi. The persistence space in multidimensional persistent homology. In *Proc. DGCI, LNCS vol. 7749*, pages 180–191, 2013. DOI: 10.1007/978-3-642-37067-0_16. 85

[44] D.-Y. Chen, X.-P. Tian, Y.-T. Shen, and M. Ouhyoung. On visual similarity based 3D model retrieval. *Computer Graphics Forum*, 22(3):223–232, 2003. DOI: 10.1111/1467-8659.00669. 93

[45] X. Chen, A. Golovinskiy, and T. Funkhouser. A benchmark for 3D mesh segmentation. *ACM Transactions on Graphics (Proc. SIGGRAPH)*, 28(3), August 2009. DOI: 10.1145/1531326.1531379. 101

[46] Y.-J. Chiang, T. Lenz, X. Lu, and G. Rote. Simple and optimal output-sensitive construction of contour trees using monotone paths. *Computational Geometry: Theory and Applications*, 30:165–195, 2005. DOI: 10.1016/j.comgeo.2004.05.002. 70

[47] W. J Cody, G. Meinardus, and R. S Varga. Chebyshev rational approximations to e^{-x} in $[0, +\infty)$ and applications to heat-conduction problems. *Journal of Approximation Theory*, 2(1):50 – 65, 1969. DOI: 10.1016/0021-9045(69)90030-6. 40

[48] K. Cole-McLaughlin, H. Edelsbrunner, J. Harer, V. Natarajan, and V. Pascucci. Loops in Reeb graphs of 2-manifolds. In *SCG '03: Proceedings of the 19th Annual Symposium on Computational Geometry*, pages 344–350, New York, NY, USA, 2003. ACM Press. DOI: 10.1145/777792.777844. 70, 71, 72

[49] L. Comic and L. De Floriani. Dimension-independent simplification and refinement of Morse complexes. *Graphical Models*, 73(5):261–285, 2011. DOI: 10.1016/j.gmod.2011.05.001. 78

[50] T. H. Cormen, C. E. Leiserson, and R. L. Rivest. *Introduction to Algorithms*. The MIT Press, Cambridge, MA, 1994. 27

[51] C. Cortes, M. Mohri, and A. Rostamizadeh. Learning non-linear combinations of kernels. In Y. Bengio, D. Schuurmans, J. Lafferty, C. K. I. Williams, and A. Culotta, editors, *Advances in Neural Information Processing Systems 22*, pages 396–404. Curran Associates, Inc., Vancouver, British Columbia, Canada, 2009. 74

[52] J. Cox, D. B. Karron, and N. Ferdous. Topological zone organization of scalar volume data. *Journal of Mathematical Imaging and Vision*, 18:95–117, 2003. DOI: 10.1023/A:1022113114311. 70

[53] M. d'Amico, P. Frosini, and C. Landi. Using matching distance in size theory: A survey. *International Journal of Imaging Systems and Technology*, 16(5):154–161, 2006. DOI: 10.1002/ima.20076. 49

[54] E. Danovaro, L. De Floriani, P. Magillo, M. M. Mesmoudi, and E. Puppo. Morphology-driven simplification and multi-resolution modeling of terrains. In E. Hoel and P. Rigaux, editors, *ACM-GIS 2003: Proceedings of the* 11th *International Symposium on Advances in Geographic Information Systems*, pages 63–70. ACM Press, November 2003. DOI: 10.1145/956676. 77

[55] E. Danovaro, L. De Floriani, and M. M. Mesmoudi. Topological analysis and characterization of discrete scalar fields. In T. Asano, R. Klette, and C. Ronse, editors, *Theoretical Foundations of Computer Vision, Geometry, Morphology, and Computational Imaging*, volume 2616 of *Lecture Notes in Computer Science*, pages 386–402. Springer Verlag, 2003. 77

[56] C. J. A. Delfinado and H. Edelsbrunner. An incremental algorithm for Betti numbers of simplicial complexes on the 3-sphere. *Computer Aided Geometric Design*, 12:771–784, 1995. DOI: 10.1016/0167-8396(95)00016-Y. 58

[57] T. K. Dey, F. Fan, and Y. Wang. An efficient computation of handle and tunnel loops via Reeb graphs. *ACM Transactions on Graphics*, 32(4):32:1–32:10, July 2013. DOI: 10.1145/2461912.2462017. 59

[58] T. K. Dey and S. Guha. Computing homology groups for simplicial complexes in $\mathbb{R}^3$. *Journal of the ACM*, 45:266–287, Mar 1998. DOI: 10.1145/274787.274810. 59

[59] T. K. Dey, A. N. Hirani, and B. Krishnamoorthy. Optimal homologous cycles, total unimodularity, and linear programming. In *Proceedings of the 42nd ACM symposium on Theory of computing*, STOC '10, pages 221–230, New York, NY, USA, 2010. ACM. DOI: 10.1145/1806689.1806721. xiii, 59

[60] T. K. Dey, K. Li, J. Sun, and D. Cohen-Steiner. Computing geometry-aware handle and tunnel loops in 3D models. *ACM Transactions on Graphics*, 27(3):45:1–45:9, August 2008. DOI: 10.1145/1360612.1360644. 59

[61] M. M. Deza and E. Deza. *Encyclopedia of Distances*. Springer Berlin Heidelberg, 2009. DOI: 10.1007/978-3-642-00234-2. 14

[62] M. P. do Carmo. *Differential Geometry of Curves and Surfaces*. Cambridge University Press, 1976. 16

[63] H. Doraiswamy and V. Natarajan. Efficient algorithms for computing Reeb graphs. *Comput. Geom. Theory Appl.*, 42(6-7):606–616, August 2009. DOI: 10.1016/j.comgeo.2008.12.003. 71, 72

[64] I. L. Dryden and K.V. Mardia. *Statistical shape analysis*. John Wiley & Sons New York, 1998. 87

[65] H. Edelsbrunner. Modeling with simplicial complexes. In *Proc. 6th Canadian Conf. Comput. Geom.*, pages 36–44, 1994. 56

[66] H. Edelsbrunner and J. Harer. Jacobi sets of multiple Morse functions. In F. Cucker, R. DeVore, P. Olver, and E. Sueli, editors, *Foundations in Computational Mathematics*, pages 37–57. Cambridge univ. Press, 2002. 64

[67] H. Edelsbrunner and J. Harer. *Computational Topology: An Introduction*. Amer. Math. Soc., 2010. 84

[68] H. Edelsbrunner, J. Harer, A. Mascarenhas, and V. Pascucci. Time-varying Reeb graphs for continuous space-time data. In *SCG '04: Proceedings of the 20th Annual Symposium on Computational Geometry*, pages 366–372, New York, NY, USA, 2004. ACM Press. DOI: 10.1145/997817.997872. 72

[69] H. Edelsbrunner, J. Harer, V. Natarajan, and V. Pascucci. Morse-Smale complexes for piecewise linear 3-manifolds. In *SCG '03: Proceedings of the 19th Annual Symposium on Computational Geometry*, pages 361–370. ACM Press, 2003. DOI: 10.1145/777792.777846. 77

[70] H. Edelsbrunner, J. Harer, and A. Zomorodian. Hierarchical Morse complexes for piecewise linear 2-manifolds. In *SCG '01: Proceedings of the 17th Annual Symposium on Computational Geometry*, pages 70–79, New York, NY, USA, 2001. ACM Press. DOI: 10.1145/378583.378626. 77, 78

[71] H. Edelsbrunner, D. Letscher, and A. Zomorodian. Topological persistence and simplification. In *IEEE Foundations of Computer Science, Proceedings 41st Annual Symposium*, pages 454–463, 2000. DOI: 10.1007/s00454-002-2885-2. 81

[72] H. Edelsbrunner, D. Letscher, and A. Zomorodian. Topological persistence and simplification. *Discrete Computational Geometry*, 28:511–533, 2002. DOI: 10.1007/s00454-002-2885-2. 81

[73] H. Edelsbrunner and E. P. Mücke. Simulation of Simplicity: A technique to cope with degenerate cases in geometric algorithms. *ACM Transactions on Graphics*, 9(1):66–104, 1990. DOI: 10.1145/77635.77639. 64

[74] C. Ehresmann and G. Reeb. Sur les champs d'éléments de contact de dimension p complètement intégrable dans une variété continuèment differentiable v_n. *Comptes Rendu Hebdomadaires des Séances de l'Académie des Sciences*, 218:955–957, 1944. 69

[75] J. Erickson and S. Har-Peled. Optimally cutting a surface into a disk. *Discrete Computational Geometry*, 31(1):37–59, 2004. DOI: 10.1007/s00454-003-2948-z. 59

[76] M. Ester, H.-P. Kriegel, J. Sander, and X. Xu. A density-based algorithm for discovering clusters in large spatial databases with noise. In *KDD*, pages 226–231. AAAI Press, 1996. 91

[77] F. Fedele, G. Gallego, A. Yezzi, A. Benetazzo, L. Cavaleri, M. Sclavo, and M. Bastianini. Euler characteristics of oceanic sea states. *Mathematics and Computers in Simulation*, 82(6):1102 – 1111, 2012. DOI: 10.1016/j.matcom.2011.05.009. 58

[78] J. D. Foley, A. van Dam, S. K. Feiner, and J. F. Hughes. *Computer graphics: principles and practice (2nd ed.)*. Addison-Wesley Longman Publishing Co., Inc., Boston, MA, USA, 1990. 56

[79] A. Fomenko. *Visual Geometry and Topology*. Springer Verlag, 1995. DOI: 10.1007/978-3-642-76235-2. 62

[80] A. Fomenko and T. L. Kunii. *Topological Modelling for Visualization*. Springer Verlag, 1997. DOI: 10.1007/978-4-431-66956-2. 18, 62

[81] P. Frosini. A distance for similarity classes of submanifolds of a Euclidean space. *Bulletin of the Australian Mathematical Society*, 42:407–416, 1990. DOI: 10.1017/S0004972700028574. 81

[82] P. Frosini. Measuring shapes by size functions. In David P. Casasent, editor, *Intelligent Robots and Computer Vision X: Algorithms and Techniques, Proceedings of SPIE*, volume 1607, pages 122–133, Boston, MA, 1991. 51

[83] P. Frosini and C. Landi. Size functions and formal series. *Applicable Algebra in Engineering, Communication and Computing*, 12:327–349, 2001. DOI: 10.1007/s002000100078. 84

[84] T. Funkhouser, P. Min, M. Kazhdan, J. Chen, A. Halderman, D. Dobkin, and D. Jacobs. A Search Engine for 3D Models. *ACM Transactions on Graphics*, 22(1):83–105, 2003. DOI: 10.1145/588272.588279. 72, 93

[85] J. Gamble and H. Giseon. Exploring uses of persistent homology for statistical analysis of landmark-based shape data. *Journal of Multivariate Analysis*, 101:2184–2199, 2010. DOI: 10.1016/j.jmva.2010.04.016. 73, 86

[86] T. D. Gatzke. Estimating curvature on triangular meshes. *International Journal of Shape Modeling*, 12(1):pp. 1–28, 2006. DOI: 10.1142/S0218654306000810. 31

[87] C. Giertsen, A. Halvorsen, and P. R. Flood. Graph-directed modelling from serial sections. *The Visual Computer*, 6:284–290, 1990. DOI: 10.1007/BF01900750. 70

[88] R. Gonzalez-Diaz, A. Ion, M. Iglesias-Ham, and W. G. Kropatsch. Invariant representative cocycles of cohomology generators using irregular graph pyramids. *Computer Vision and Image Understanding*, 7(115):1011–1022, 2011. DOI: 10.1016/j.cviu.2010.12.009. 85

[89] A. Gramain. *Topologie des surfaces*. Presses Universitaires de France, 1971. 65

[90] M. Greenberg and J. R. Harper. *Algebraic topology: A first course*. Addison-Wesley, 1981. 55, 67

[91] E. Grinspun, M. Desbrun, K. Polthier, P. Schröder, and A. Stern. Discrete differential geometry: An applied introduction. In *ACM SIGGRAPH 2005 Courses*, SIGGRAPH '05, New York, NY, USA, 2005. ACM. DOI: 10.1145/1198555.1198660. 36

[92] M. Gromov. *Metric Structures for Riemannian and Non-Riemannian Spaces Couverture*. Springer, 1999. 49

[93] X. Gu, S. J. Gortler, and H. Hoppe. Geometry images. *ACM Transactions on Graphics*, 21(3):355–361, July 2002. DOI: 10.1145/566654.566589. 59

[94] V. Guillemin and A. Pollack. *Differential Topology*. Englewood Cliffs, NJ: Prentice-Hall, 1974. 18, 19, 62

[95] D. Günther, J. Reininghaus, H. P. Seidel, and T. Weinkauf. Notes on the simplification of the Morse-Smale complex. In *Proceedings TopoInVis*, Davis, USA, March 2013. DOI: 10.1007/978-3-319-04099-8_9. 78

[96] A. Gyulassy, V. Natarajan, V. Pascucci, P.-T. Bremer, , and B. Hamann. Topology-based simplification for feature extraction from 3D scalar fields. In *VIS '05: Proceedings of the IEEE Visualization 2005*, pages 275–280. IEEE Computer Society Press, 2005. DOI: 10.1109/VISUAL.2005.1532839. 78

[97] A. Gyulassy, V. Natarajan, V. Pascucci, and B. Hamann. Efficient computation of Morse-Smale complexes for three-dimensional scalar functions. *IEEE Transactions on Visualization and Computer Graphics*, 13(6):1440–1447, 2007. DOI: 10.1109/TVCG.2007.70552. 77, 78

[98] W. Harvey, Y. Wang, and R. Wenger. A randomized $O(M \log M)$ time algorithm for computing Reeb graphs of arbitrary simplicial complexes. In *Proceedings of the Twenty-sixth Annual Symposium on Computational Geometry*, SoCG '10, pages 267–276, New York, NY, USA, 2010. ACM. DOI: 10.1145/1810959.1811005. 71, 72

[99] A. Hatcher. *Algebraic Topology*. Cambridge University Press, 2001. 53

[100] M. Hein, J.-Y. Audibert, and U. von Luxburg. From graphs to manifolds-weak and strong pointwise consistency of graph Laplacians. volume 3559 of *Lecture Notes in Computer Science*, pages 470–485. Springer, 2005. DOI: 10.1007/11503415_32. 37

[101] J. Helman and L. Hesselink. Representation and display of vector field topology in fluid flow data sets. *Computer*, pages 27–36, 1989. DOI: 10.1109/2.35197. 79

[102] M. Hilaga, Y. Shinagawa, T. Kohmura, and T. L. Kunii. Topology matching for fully automatic similarity estimation of 3D shapes. In *SIGGRAPH '01: Proceedings of the 28th Annual Conference on Computer Graphics and Interactive Techniques*, pages 203–212, Los Angeles, CA, August 2001. ACM Press. DOI: 10.1145/383259.383282. 27, 28, 29, 70

[103] A. N. Hirani. *Discrete Exterior Calculus.* PhD thesis, California Institute of Technology, May 2003. 5

[104] M. W. Hirsch. *Differential Topology.* Springer, 1997. 61

[105] R. Hu, L. Fan, and L. Liu. Co-segmentation of 3D shapes via subspace clustering. In *SGP'12: Proceedings of the 2012 Eurographics Symposium on Geometry Processing*, pages 1703–1713, 2012. DOI: 10.1111/j.1467-8659.2012.03175.x. 90

[106] S.-S. Huang, A. Shamir, C.-H. Shen, H. Zhang, A. Sheffer, S.-M. Hu, and D. Cohen-Or. Qualitative organization of collections of shapes via quartet analysis. *ACM Transactions on Graphics*, 32(4):71:1–71:10, July 2013. DOI: 10.1145/2461912.2461954. xiv, 92, 93

[107] J. A. Iglesias and R. Kimmel. Schrödinger diffusion for shape analysis with texture. In *ECCV*, pages 123–132, 2012. DOI: 10.1007/978-3-642-33863-2_13. 90

[108] E. Kalogerakis, A. Hertzmann, and K. Singh. Learning 3D mesh segmentation and labeling. *ACM Trans. Graph.*, 29:102:1–102:12, July 2010. DOI: 10.1145/1778765.1778839. 90

[109] H. Kashima, K. Tsuda, and A. Inokuchi. Marginalized kernels between labeled graphs. In 20th *Int. Conf. on Machine Learning*, pages 321–328, Washington, DC, 2003. AAAI Press. 73

[110] S. Katz and A. Tal. Hierarchical mesh decomposition using fuzzy clustering and cuts. *ACM. Transactions on Graphics*, 22:954–961, July 2003. DOI: 10.1145/882262.882369. 29

[111] D. Kendall. The diffusion of shape. *Advances in Applied Probability*, 9:428–430, 1977. DOI: 10.2307/1426091. 2

[112] V. G. Kim, W. Li, N. J. Mitra, S. Chaudhuri, S. DiVerdi, and T. Funkhouser. Learning part-based templates from large collections of 3D shapes. *ACM Transactions on Graphics*, 32(4):70:1–70:12, July 2013. DOI: 10.1145/2461912.2461933. 92

[113] V. G. Kim, W. Li, N. J. Mitra, S. DiVerdi, and T. Funkhouser. Exploring collections of 3D models using fuzzy correspondences. *ACM Transactions on Graphics*, 31(4):54:1–54:11, July 2012. DOI: 10.1145/2185520.2185550. 92

[114] V. G. Kim, Y. Lipman, X. Chen, and T. A. Funkhouser. Möbius transformations for global intrinsic symmetry analysis. *Computer Graphics Forum*, 29(5):1689–1700, 2010. DOI: 10.1111/j.1467-8659.2010.01778.x. 29, 48

[115] R. Kimmel and J. A. Sethian. Computing geodesic paths on manifolds. In *Proc. Natl. Acad. Sci. USA*, pages 8431–8435, 1998. DOI: 10.1073/pnas.95.15.8431. 28

[116] R. Klette and A. Rosenfeld. *Digital Geometry: Geometric Methods for Digital Picture Analysis*. Morgan Kaufmann Series in Computer Graphics and Geometric Modeling. Morgan Kaufmann, 2004. 56

[117] J. J. Koenderink. *Solid shape*. MIT Press, Cambridge, MA, USA, 1990. 3

[118] A. N. Kolmogorov and S. V. Fomin. *Introductory real analysis*. 1975, Dover. 19, 21

[119] R. Kondor and T. Jebara. A kernel between sets of vectors. In *Int. Conf. on Machine Learning*, Washington, DC, USA, 2003. AAAI Press. 73

[120] A. Kovnatsky, M. M. Bronstein, A. M. Bronstein, K. Glashoff, and R. Kimmel. Coupled quasi-harmonic bases. *Computer Graphics Forum*, 32(2pt4):439–448, 2013. DOI: 10.1111/cgf.12064. 90

[121] A. Kovnatsky, D. Raviv, M. M. Bronstein, A. M. Bronstein, and R. Kimmel. Geometric and photometric data fusion in non-rigid shape analysis. *Numerical Mathematics: Theory, Methods and Applications (NM-TMA)*, 6(1):199–222, 2013. 89, 90

[122] H. Laga, M. Mortara, and M. Spagnuolo. Geometry and context for semantic correspondences and functionality recognition in man-made 3D shapes. *ACM Trans. Graph.*, 32(5):150:1–150:16, October 2013. DOI: 10.1145/2516971.2516975. 89, 94

[123] S. Lang. *Real Analysis*. Addison-Wesley Publishing Company, 1983. 19, 21

[124] G. Leifman, E. Shtrom, and A. Tal. Surface regions of interest for viewpoint selection. In *Computer Vision and Pattern Recognition (CVPR), 2012 IEEE Conference on*, pages 414–421, 2012. DOI: 10.1109/CVPR.2012.6247703. 90

[125] B. Levy. Laplace-Beltrami eigenfunctions towards an algorithm that "understands" geometry. In *Proceedings of the IEEE International Conference on Shape Modeling and Applications 2006*, SMI '06, pages 13–, Washington, DC, USA, 2006. IEEE Computer Society. DOI: 10.1109/SMI.2006.21. xiii, 38, 39

[126] B. Lévy and B. Vallet. Spectral geometry processing with manifold harmonics. *Computer Graphics Forum*, 2(27), 2008. DOI: 10.1111/j.1467-8659.2008.01122.x. 36

[127] Z. Lian, A. Godil, B. Bustos, M. Daoudi, J. Hermans, S. Kawamura, Y. Kurita, G. Lavoué, H. Van Nguyen, R. Ohbuchi, Y. Ohkita, Y. Ohishi, F. Porikli, M. Reuter, I. Sipiran, D. Smeets, P. Suetens, H. Tabia, and D. Vandermeulen. A comparison of methods for non-rigid 3D shape retrieval. *Pattern Recognition*, 46(1):449–461, January 2013. DOI: 10.1016/j.patcog.2012.07.014. 90

[128] Y. Lipman and T. Funkhouser. Möbius voting for surface correspondence. *ACM Trans. Graph.*, 28(3):72:1–72:12, July 2009. DOI: 10.1145/1531326.1531378. xiii, 46, 47, 48

[129] Y.-J. Liu, Y.-F. Zheng, L. Lv, Y.-M. Xuan, and X.-L. Fu. 3D model retrieval based on color + geometry signatures. *The Visual Computer*, 28(1):75–86, 2012. DOI: 10.1007/s00371-011-0605-8. 89

[130] H. Maehara. Why is $\mathbb{P}^2$ Not Embeddable in $\mathbb{R}^3$? *The American Mathematical Monthly*, 100(9):pp. 862–864, 1993. DOI: 10.2307/2324664. 56

[131] P. Magillo, E. Danovaro, L. De Floriani, L. Papaleo, and M. Vitali. Extracting terrain morphology: A new algorithm and a comparative evaluation. In *Proceedings of the 2nd International Conference on Computer Graphics Theory and Applications*, March 8–11 '2007. 77

[132] T. Maldonado. *Reale e Virtuale*. Feltrinelli, 1994. 2

[133] A. Mangan and R. Whitaker. Partitioning 3D surface meshes using watershed segmentation. *IEEE Transaction on Visualization and Computer Graphics*, 5(4):308–321, 1999. DOI: 10.1109/2945.817348. 77

[134] M. Mantyla. *Introduction to Solid Modeling*. WH Freeman & Co. New York, NY, USA, 1988. 7, 53

[135] W. Massey. *Algebraic Topology: An Introduction*. Brace & World, Inc, 1967. 53

[136] J. C. Maxwell. On Hills and Dales. *The London, Edinburgh and Dublin Philosophical*, 40(269):421–425, 1870. 75

[137] D. S. Meek and D. J. Walton. On surface normal and gaussian curvature approximations given data sampled from a smooth surface. *Comput. Aided Geom. Des.*, 17(6):521–543, July 2000. DOI: 10.1016/S0167-8396(00)00006-6. 31

[138] F. Mémoli. Gromov-Wasserstein distances and the metric approach to object matching. *Foundations of Computational Mathematics*, 11(4):417–487, 2011. DOI: 10.1007/s10208-011-9093-5. 50

[139] F. Mémoli. Some properties of Gromov-Hausdorff distances. *Discrete & Computational Geometry*, pages 1–25, 2012. DOI: 10.1007/s00454-012-9406-8. 50

[140] F. Mémoli and G. Sapiro. A theoretical and computational framework for isometry invariant recognition of point cloud data. *Foundations of Computational Mathematics*, 5(3):313–347, July 2005. DOI: 10.1007/s10208-004-0145-y. 50

[141] J. Milnor. *Morse Theory*. Princeton University Press, New Jersey, 1963. 61

[142] J. Milnor. *Lectures on h-cobordism*. Princeton University Press, 1965. 65

[143] J. S. B. Mitchell, D. M. Mount, and C. H. Papadimitriou. The discrete geodesic problem. *SIAM J. Comput.*, 16(4):647–668, August 1987. DOI: 10.1137/0216045. 27

[144] C. Moler and C. Van Loan. Nineteen dubious ways to compute the exponential of a matrix, twenty-five years later. *SIAM Review*, 45(1):3–49, 2003. DOI: 10.1137/S00361445024180. 40

[145] M. Mortara, G. Patané, M. Spagnuolo, B. Falcidieno, and J. Rossignac. Blowing bubbles for multi-scale analysis and decomposition of triangle meshes. *Algorithmica*, 38(2):227–248, 2003. DOI: 10.1007/s00453-003-1051-4. xiii, 32, 33

[146] M. Mortara and M. Spagnuolo. Semantics-driven best view of 3D shapes. *Computers & Graphics*, 33(3):280 – 290, 2009. IEEE International Conference on Shape Modelling and Applications 2009. DOI: 10.1016/j.cag.2009.03.003. 90

[147] M. E. Mortenson. *Geometric Modeling*. John Wiley & Sons, 1986. 53

[148] H. Mueller, P. Clough, T. Deselaers, and B. Caputo. *ImageCLEF: Experimental Evaluation in Visual Information Retrieval*. Springer Publishing Company, Incorporated, 1st edition, 2010. DOI: 10.1007/978-3-642-15181-1. 100

[149] J. Munkres. *Topology*. Prentice Hall, 2000. 18, 19, 53, 54, 56, 58

[150] L.R. Nackman. Two-dimensional critical point configuration graphs. *IEEE Transactions on Pattern Analysis and Machine Intelligence*, 6(4):442–450, 1984. DOI: 10.1109/T-PAMI.1984.4767549. 7

[151] X. Ni, M. Garland, and J. C. Hart. Fair Morse functions for extracting the topological structure of a surface mesh. *ACM Transactions on Graphics*, 23(3):613–622, 2004. DOI: 10.1145/1015706.1015769. 77

[152] R. Osada, T. A. Funkhouser, B. Chazelle, and D. P. Dobkin. Shape distributions. *ACM Transactions on Graphics*, 21(4):807–832, 2002. DOI: 10.1145/571647.571648. 94

[153] P. Over, G. Awad, M. Michel, J. Fiscus, G. Sanders, B. Shaw, W. Kraaij, A. F. Smeaton, and G. Qu enot. TRECVID 2012 – an overview of the goals, tasks, data, evaluation mechanisms and metrics. In *Proceedings of TRECVID 2012*. NIST, USA, 2012. 100

[154] M. Ovsjanikov, J. Sun, and L. Guibas. Global intrinsic symmetries of shapes. In *Proceedings of the Symposium on Geometry Processing*, SGP '08, pages 1341–1348, Aire-la-Ville, Switzerland, Switzerland, 2008. Eurographics Association. DOI: 10.1111/j.1467-8659.2008.01273.x. 38

[155] D. L. Page. *Part Decomposition of 3D Surfaces*. PhD thesis, University of Tennessee, Knoxville, May 2003. 77

[156] J. Palis and W. De Melo. *Geometric Theory of Dynamical Systems: An Introduction*. Springer-Verlag, 1982. DOI: 10.1007/978-1-4612-5703-5. 75, 76

[157] S. Parsa. A deterministic $O(M \log M)$ time algorithm for the Reeb graph. In *Proceedings of the Twenty-eighth Annual Symposium on Computational Geometry*, SoCG '12, pages 269–276, New York, NY, USA, 2012. ACM. DOI: 10.1145/2261250.2261289. 71, 72

[158] V. Pascucci. Topology diagrams of scalar fields in scientific visualization. In S. Rana, editor, *Topological Data Structures for Surfaces*, pages 121–129. John Wiley & Sons Ltd, 2004. DOI: 10.1002/0470020288. 77

[159] V. Pascucci, G. Scorzelli, P.-T. Bremer, and A. Mascarenhas. Robust on-line computation of Reeb graphs: Simplicity and speed. *ACM Transactions on Graphics*, 26(3), July 2007. DOI: 10.1145/1276377.1276449. 70, 72

[160] G. Pasqualotto, P. Zanuttigh, and G. M. Cortelazzo. Combining color and shape descriptors for 3D model retrieval. *Signal Process-Image*, 28(6):608 – 623, 2013. DOI: 10.1016/j.image.2013.01.004. 89

[161] G. Patanè and M. Spagnuolo. Heat diffusion kernel and distance on surface meshes and point sets. *Computers & Graphics*, 37(6):676 – 686, 2013. DOI: 10.1016/j.cag.2013.05.019. 40

[162] G. Peyré, M. Pechaud, R. Keriven, and L. D. Cohen. Geodesic methods in computer vision and graphics. *Foundations and Trends in Computer Graphics and Vision*, 5(3-4):197–397, 2010. DOI: 10.1561/0600000029. 27, 28

[163] J. L. Pfaltz. Surface networks. *Geographical Analysis*, 8:77–93, 1976. 77

[164] K. Polthier and M. Schmies. Straightest geodesics on polyhedral surfaces. In H.C. Hege and K. Polthier, editors, *Mathematical Visualization*. Springer Verlag, 1998. DOI: 10.1007/978-3-662-03567-2. 31

[165] S. Rana, editor. *Topological Data Structures for Surfaces: An Introduction for Geographical Information Science*. John Wiley & Sons, Europe, London, 2004. DOI: 10.1002/0470020288. 77

[166] G. Reeb. Sur les points singuliers d'une forme de Pfaff complètement intégrable ou d'une fonction numérique. *Comptes Rendus Hebdomadaires des Séances de l'Académie des Sciences*, 222:847–849, 1946. 69

[167] A. Requicha. Representations of rigid solids: Theory, methods and systems. *ACM Computing Surveys*, 12(4):437–464, 1980. DOI: 10.1145/356827.356833. 7, 19

[168] M. Reuter. *Laplace Spectra for Shape Recognition*. Books on Demand, ISBN 3-8334-5071-1, 2006. 36

[169] M. Reuter, S. Biasotti, D. Giorgi, G. Patanè, and M. Spagnuolo. Discrete Laplace-Beltrami operators for shape analysis and segmentation. *Comput. Graph.*, 33(3):381–390, June 2009. DOI: 10.1016/j.cag.2009.03.005. xiii, 37, 38

[170] V. Robins. Towards computing homology from finite approximations. In W.W. Comfort, R. Heckmann, R.D. Kopperman, and L. Narici, editors, 14th *Summer Conference on General Topology and its Applications*, volume 24 of *Topology Proceedings*, pages 503–532, 1999. 81

[171] S. Rosenberg. *The Laplacian on a Riemannian Manifold*. Cambridge University Press, 1997. DOI: 10.1017/CBO9780511623783. 36

[172] C. R. Ruiz, R. Cabredo, L. J. Monteverde, and Z. Huang. Combining Shape and Color for Retrieval of 3D Models. In *INC, IMS and IDC, 2009. NCM '09. Fifth International Joint Conference on*, pages 1295–1300, 2009. DOI: 10.1109/NCM.2009.140. 89

[173] H. Sato. *Algebraic Topology: An Intuitive Approach*. Translations in Modern Mathematics, Iwanami Series in Modern Mathematics. American Mathematical Society, Providence, Rhone Island, 1999. 53

[174] B. Schneider and J. Wood. Construction of metric surface networks from raster-based DEMs. In S. Rana, editor, *Topological Data Structures for Surfaces: An Introduction for Geographical Information Science*, pages 53–70. John Wiley & Sons Ltd, 2004. DOI: 10.1002/0470020288. 77

[175] J. A. Sethian. A fast marching level set method for monotonically advancing fronts. In *Proc. Nat. Acad. Sci*, pages 1591–1595, 1995. 28

[176] J. A. Sethian. *Level Set Methods and Fast Marching Methods: Evolving Interfaces in Computational Geometry, Fluid Mechanics, Computer Vision, and Materials Science*. Cambridge University Press, 1999. 28

[177] A. Shamir. A survey on mesh segmentation techniques. *Computer Graphics Forum*, 27(6):1539–1556, 2008. DOI: 10.1111/j.1467-8659.2007.01103.x. 29, 31

[178] L. Shapira, S. Shalom, A. Shamir, D. Cohen-Or, and H. Zhang. Contextual part analogies in 3D objects. *Int. J. of Comput. Vision*, 89(2-3):309–326, 2010. DOI: 10.1007/s11263-009-0279-0. 29

[179] L. Shapira, A. Shamir, and D. Cohen-Or. Consistent mesh partitioning and skeletonisation using the shape diameter function. *Vis. Comput.*, 24(4):249–259, March 2008. DOI: 10.1007/s00371-007-0197-5. 93

[180] D. Shattuck and R. Leahy. Automated graph based analysis and correction of cortical volume topology. *IEEE Transactions on Medical Imaging*, 20:1167–1177, 2001. DOI: 10.1109/42.963819. 72

[181] Y. Shinagawa, T. L. Kunii, and Y. L. Kergosien. Surface coding based on Morse theory. *IEEE Computer Graphics and Applications*, 11:66–78, 1991. DOI: 10.1109/38.90568. 70, 72

[182] O. Sidi, O. van Kaick, Y. Kleiman, H. Zhang, and D. Cohen-Or. Unsupervised co-segmentation of a set of shapes via descriptor-space spectral clustering. *ACM Transactions on Graphics*, 30(6):126:1–126:10, December 2011. DOI: 10.1145/2070781.2024160. 102

[183] A. Singer. From graph to manifold Laplacian: The convergence rate. *Applied and Computational Harmonic Analysis*, 21(1):128–134, July 2006. DOI: 10.1016/j.acha.2006.03.004. 37

[184] A. F. Smeaton, P. Over, and W. Kraaij. Evaluation campaigns and TRECVid. In *MIR '06: Proceedings of the 8th ACM International Workshop on Multimedia Information Retrieval*, pages 321–330, New York, NY, USA, 2006. ACM Press. DOI: 10.1145/1178677.1178722. 100

[185] A. F. Smeaton, P. Over, and W. Kraaij. High-Level Feature Detection from Video in TRECVid: a 5-Year Retrospective of Achievements. In Ajay Divakaran, editor, *Multimedia Content Analysis, Theory and Applications*, pages 151–174. Springer Verlag, Berlin, 2009. DOI: 10.1007/978-0-387-76569-3. 100

[186] S. Snir and S. Rao. Quartets maxcut: A divide and conquer quartets algorithm. *IEEE/ACM Transactions on Computational Biology and Bioinformatics*, 7(4):704–718, 2010. DOI: 10.1109/TCBB.2008.133. 93

[187] O. Sorkine. Differential representations for mesh processing. *Computer Graphics Forum*, 25(4):789–807, 2006. DOI: 10.1111/j.1467-8659.2006.00999.x. 35

[188] E. H. Spanier. *Algebraic Topology*. McGraw Hill, 1966. 56

[189] A. Spira and R. Kimmel. An efficient solution to the Eikonal equation on parametric manifolds. In *INTERPHASE 2003 meeting, Isaac Newton Institute for Mathematical Sciences, 2003 Preprints, Preprint no. NI03045-CPD*, volume 3, pages 315–327, 2003. DOI: 10.4171/IFB/102. 28

[190] J. Sun, M. Ovsjanikov, and L. Guibas. A concise and provably informative multi-scale signature based on heat diffusion. *Computer Graphics Forum*, 28(5):1383–1392, 2009. DOI: 10.1111/j.1467-8659.2009.01515.x. 40, 41

[191] V. Surazhsky, T. Surazhsky, D. Kirsanov, S. J. Gortler, and H. Hoppe. Fast exact and approximate geodesics on meshes. *ACM Trans. Graph.*, 24(3):553–560, July 2005. DOI: 10.1145/1073204.1073228. 27

[192] S. Takahashi, G. M. Nielson, Y. Takeshima, and I. Fujishiro. Topological volume skeletonization using adaptive tetrahedralization. In *Geometric Modelling and Processing 2004*, pages 227–236, 2004. DOI: 10.1109/GMAP.2004.1290044. 70

[193] G. Taubin. A signal processing approach to fair surface design. In *Proceedings of the 22Nd Annual Conference on Computer Graphics and Interactive Techniques*, SIGGRAPH '95, pages 351–358, New York, NY, USA, 1995. ACM. DOI: 10.1145/218380.218473. 35

[194] H. Theisel, C. Roessl, and T. Weinkau. Morphological representations of vector fields. In L. De Floriani and M. Spagnuolo, editors, *Shape Analysis and Structuring*. 2007. 79

[195] D. W. Thompson, editor. *On Growth and Form*. Cambridge University Press, Cambridge, MA, 1942. xiii, 2, 3

[196] J. Tierny, A. Gyulassy, E. Simon, and V. Pascucci. Loop surgery for volumetric meshes: Reeb graphs reduced to contour trees. *IEEE Transactions on Visualization and Computer Graphics*, 15(6):1177–1184, November 2009. DOI: 10.1109/TVCG.2009.163. 72

[197] J. Tierny, J.-P. Vandeborre, and M. Daoudi. Partial 3D shape retrieval by Reeb pattern unfolding. *Comput. Graph. Forum*, 28(1):41–55, 2009. DOI: 10.1111/j.1467-8659.2008.01190.x. xiv, 29, 74

[198] A. N. Tikhonov and A. A. Samarski. *Equations of Mathematical Physics*. Dover, 1990. 36

[199] F. Tombari, S. Salti, and L. Di Stefano. A combined texture-shape descriptor for enhanced 3D feature matching. In *ICIP*, pages 809–812, 2011. DOI: 10.1109/ICIP.2011.6116679. 89

[200] F. Tombari, S. Salti, and L. Di Stefano. Performance evaluation of 3D keypoint detectors. *International Journal of Computer Vision*, 102(1-3):198–220, 2013. DOI: 10.1007/s11263-012-0545-4. 100

[201] A. Torsello. *Matching hierarchical structures for shape recognition, Ph.D. Thesis.* University of York, UK, 2004. 73

[202] T. Tung and F. Schmitt. The Augmented Multiresolution Reeb Graph approach for content-based retrieval of 3D shapes. *International Journal of Shape Modelling*, 11(1):91–120, June 2005. DOI: 10.1142/S0218654305000748. 70

[203] G. Vegter. *Computational Topology*, pages 517–536. CRC Press, 1997. 56, 59

[204] G. Vegter and C. K. Yap. Computational complexity of combinatorial surfaces. In *SCG '90: Proceedings of the 6th Annual Symposium on Computational Geometry*, pages 102–111, New York, NY, USA, 1990. ACM Press. DOI: 10.1145/98524.98546. 58

[205] R. C. Veltkamp, R. Ruijsenaars, M. Spagnuolo, R. van Zwol, and F. ter Haar. SHREC2006: 3D shape retrieval contest. Technical Report UU-CS-2006-030, Department of Information and Computing Sciences, Utrecht University, 2006. 100

[206] S. V. N. Vishwanathan, N. N. Schraudolph, R. Kondor, and K. Borgwardt. Graph kernels. *Journal of Machine Learning Research*, 11:1201–1242, 2010. 73

[207] C. Wallraven, B. Caputo, and A. Graf. Recognition with local features: the kernel recipe. In *Proceedings of the Ninth IEEE International Conference on Computer Vision - Volume 2*, ICCV '03, pages 257–, Washington, DC, USA, 2003. IEEE Computer Society. DOI: 10.1109/ICCV.2003.1238351. 73

[208] Y. Wang, S. Asafi, O. van Kaick, H. Zhang, D. Cohen-Or, and B. Chen. Active co-analysis of a set of shapes. *ACM Transactions on Graphics*, 31(6):165:1–165:10, November 2012. DOI: 10.1145/2366145.2366184. 90

[209] M. Wardetzky, S. Mathur, F. Kälberer, and E. Grinspun. Discrete Laplace operators: no free lunch. In *Proc. of the Symp. on Geometry processing*, pages 33–37, 2007. 36, 37

[210] O. Weber, Y. S. Devir, A. M. Bronstein, M. M. Bronstein, and R. Kimmel. Parallel algorithms for approximation of distance maps on parametric surfaces. *ACM Transactions on Graphics*, 27(4):104:1–104:16, November 2008. DOI: 10.1145/1409625.1409626. 28

[211] N. Werghi, Y. Xiao, and J. P. Siebert. A functional-based segmentation of human body scans in arbitrary postures. *IEEE Transactions on Systems, Man, and Cybernetics - Part B: Cybernetics*, 36(1):153–165, 2006. DOI: 10.1109/TSMCB.2005.854503. 70

[212] S. Willard. *General topology*. Addison-Wesley Publishing Company, 1970. 17

[213] A. Zaharescu, E. Boyer, and R. Horaud. Keypoints and local descriptors of scalar functions on 2D manifolds. *Int. J. Comput. Vision*, 100(1):78–98, 2012. DOI: 10.1007/s11263-012-0528-5. 89

[214] E. Zhang, K. Mischaikow, and G. Turk. Feature-based surface parameterization and texture mapping. *ACM Transactions on Graphics*, 24(1):1–27, 2005. DOI:
10.1145/1037957.1037958. 29

[215] H. Zhang, O. van Kaick, and R. Dyer. Spectral mesh processing. *Computer Graphics
Forum*, 29(6):1865–1894, 2010. DOI: 10.1111/j.1467-8659.2010.01655.x. 35

[216] A. Zomorodian and G. Carlsson. Computing persistent homology. *Discrete and Computational Geometry*, 33(2):249–274, 2005. DOI: 10.1007/s00454-004-1146-y. 85

[217] S. Biasotti, B. Falcidieno, D. Giorgi, and M. Spagnuolo. The hitchiker's guide to the galaxy
of mathematical tools for shape analysis. In *ACM SIGGRAPH 2012 Courses* (SIGGRAPH
'12), New York, NY, USA, 2012. Article 17, 33 pages. DOI: 10.1145/2343483.2343499.
3, 20, 21, 22, 45, 50

Authors' Biographies

SILVIA BIASOTTI

Silvia Biasotti is a researcher at CNR-IMATI. She graduated in Mathematics in 1998 and got a Ph.D. in Mathematics and Applications in 2004 and a Ph.D. in Information and Communication Technologies in 2008, all from the University of Genoa. She is co-author of more than 80 reviewed papers on geometric modeling, shape analysis and computational topology. Since 2005 she has been responsible for the CNR activity Topology and Homology for analyzing digital shapes, and teaches the Master course on methods of analysis of discrete surfaces and their applications at the Department of Mathematics, University of Genoa.

BIANCA FALCIDIENO

Bianca Falcidieno is a research director of the National Research Council of Italy at CNR-IMATI. She has been leading and coordinating research at international level in Computational Mathematics, Computer Graphics, 3D Media and Knowledge Technologies, strongly interacting with industrial application fields. Author of more than 200 scientific reviewed papers and books, she was the editor in chief of the *International Journal of Shape Modeling* and guest-editor of several special issues. Since 2011 she is Fellow of the EUROGRAPHICS Association and for the 80th CNR anniversary she was included in the 12 top-level female researchers in the CNR history.

DANIELA GIORGI

Daniela Giorgi graduated cum laude in Mathematics at the University of Bologna (2002), and got a Ph.D. in Computational Mathematics from the University of Padova (2006). Since then she has been a researcher at the National Research Council of Italy. She has considerable expertise in mathematics (geometry and topology), and in image and 3D shape analysis. She has authored about 40 peer-reviewed publications on computational geometry and topology for shape analysis, and has been participating in several international projects on related topics. She has been teaching BS and Master (Mathematics and its Applications) courses, and has been a lecturer at international schools.

MICHELA SPAGNUOLO

Michela Spagnuolo got a Laurea Degree cum laude in Applied Mathematics from the University of Genova and a Ph.D. in Computer Science Engineering at INSA, Lyon. She is currently a research director at CNR-IMATI. She authored more than 130 reviewed papers, edited a book on 3D shape analysis, and was a guest-editor of several special issues. She is an associate editor of *Computers & Graphics*, and *The Visual Computer*. She is a member of the steering committee of Shape Modelling International, and of the EG workshops on 3D Object Retrieval. Since 2014 she is Fellow of the EUROGRAPHICS Association. Her interests include computational topology, shape analysis, shape similarity and matching.

the United States
& Taylor Publisher Services